CONTINUOUS FLOW MANUFACTURING

MANUFACTURING ENGINEERING AND MATERIALS PROCESSING
A Series of Reference Books and Textbooks

FOUNDING EDITOR

Geoffrey Boothroyd
University of Rhode Island
Kingston, Rhode Island

ADVISORY EDITORS

Gary F. Benedict   **E. A. Elsayed**
Allied-Signal   *Rutgers University*

Fred W. Kear   **Michel Roboam**
Motorola   *Aerospatiale*

Jack Walker
McDonnell Douglas

1. Computers in Manufacturing, *U. Rembold, M. Seth, and J. S. Weinstein*
2. Cold Rolling of Steel, *William L. Roberts*
3. Strengthening of Ceramics: Treatments, Tests, and Design Applications, *Harry P. Kirchner*
4. Metal Forming: The Application of Limit Analysis, *Betzalel Avitzur*
5. Improving Productivity by Classification, Coding, and Data Base Standardization: The Key to Maximizing CAD/CAM and Group Technology, *William F. Hyde*
6. Automatic Assembly, *Geoffrey Boothroyd, Corrado Poli, and Laurence E. Murch*
7. Manufacturing Engineering Processes, *Leo Alting*
8. Modern Ceramic Engineering: Properties, Processing, and Use in Design, *David W. Richerson*
9. Interface Technology for Computer-Controlled Manufacturing Processes, *Ulrich Rembold, Karl Armbruster, and Wolfgang Ülzmann*
10. Hot Rolling of Steel, *William L. Roberts*
11. Adhesives in Manufacturing, *edited by Gerald L. Schneberger*
12. Understanding the Manufacturing Process: Key to Successful CAD/ CAM Implementation, *Joseph Harrington, Jr.*
13. Industrial Materials Science and Engineering, *edited by Lawrence E. Murr*
14. Lubricants and Lubrication in Metalworking Operations, *Elliot S. Nachtman and Serope Kalpakjian*
15. Manufacturing Engineering: An Introduction to the Basic Functions, *John P. Tanner*
16. Computer-Integrated Manufacturing Technology and Systems, *Ulrich Rembold, Christian Blume, and Ruediger Dillman*
17. Connections in Electronic Assemblies, *Anthony J. Bilotta*

18. Automation for Press Feed Operations: Applications and Economics, *Edward Walker*
19. Nontraditional Manufacturing Processes, *Gary F. Benedict*
20. Programmable Controllers for Factory Automation, *David G. Johnson*
21. Printed Circuit Assembly Manufacturing, *Fred W. Kear*
22. Manufacturing High Technology Handbook, *edited by Donatas Tijunelis and Keith E. McKee*
23. Factory Information Systems: Design and Implementation for CIM Management and Control, *John Gaylord*
24. Flat Processing of Steel, *William L. Roberts*
25. Soldering for Electronic Assemblies, *Leo P. Lambert*
26. Flexible Manufacturing Systems in Practice: Applications, Design, and Simulation, *Joseph Talavage and Roger G. Hannam*
27. Flexible Manufacturing Systems: Benefits for the Low Inventory Factory, *John E. Lenz*
28. Fundamentals of Machining and Machine Tools: Second Edition, *Geoffrey Boothroyd and Winston A. Knight*
29. Computer-Automated Process Planning for World-Class Manufacturing, *James Nolen*
30. Steel-Rolling Technology: Theory and Practice, *Vladimir B. Ginzburg*
31. Computer Integrated Electronics Manufacturing and Testing, *Jack Arabian*
32. In-Process Measurement and Control, *Stephan D. Murphy*
33. Assembly Line Design: Methodology and Applications, *We-Min Chow*
34. Robot Technology and Applications, *edited by Ulrich Rembold*
35. Mechanical Deburring and Surface Finishing Technology, *Alfred F. Scheider*
36. Manufacturing Engineering: An Introduction to the Basic Functions, Second Edition, Revised and Expanded, *John P. Tanner*
37. Assembly Automation and Product Design, *Geoffrey Boothroyd*
38. Hybrid Assemblies and Multichip Modules, *Fred W. Kear*
39. High-Quality Steel Rolling: Theory and Practice, *Vladimir B. Ginzburg*
40. Manufacturing Engineering Processes: Second Edition, Revised and Expanded, *Leo Alting*
41. Metalworking Fluids, *edited by Jerry P. Byers*
42. Coordinate Measuring Machines and Systems, *edited by John A. Bosch*
43. Arc Welding Automation, *Howard B. Cary*
44. Facilities Planning and Materials Handling: Methods and Requirements, *Vijay S. Sheth*
45. Continuous Flow Manufacturing: Quality in Design and Processes, *Pierre C. Guerindon*

Additional Volumes in Preparation

CONTINUOUS FLOW MANUFACTURING

QUALITY IN DESIGN AND PROCESSES

PIERRE C. GUERINDON

Group President, Retired
Caterpillar, Inc.
Peoria, Illinois

Taylor & Francis
Taylor & Francis Group
Boca Raton   London   New York

CRC is an imprint of the Taylor & Francis Group,
an informa business

Published in 1995 by
CRC Press
Taylor & Francis Group
6000 Broken Sound Parkway NW, Suite 300
Boca Raton, FL 33487-2742

No claim to original U.S. Government works
Printed in the United States of America on acid-free paper
10 9 8 7 6 5 4 3 2

International Standard Book Number-10: 0-8247-8655-1 (Hardcover)
International Standard Book Number-13: 978-0-8247-8655-6 (Hardcover)
Library of Congress catalog number: 95-32264

Library of Congress Cataloging-in-Publication Data

Catalog record is available from the Library of Congress

Taylor & Francis Group
is the Academic Division of Informa plc.

Visit the Taylor & Francis Web site at
http://www.taylorandfrancis.com

and the CRC Press Web site at
http://www.crcpress.com

Foreword

Determined. Aggressive. Visionary. Persevering. Impatient. Energized. Relentless. Convincing. Tenacious. These are all words that describe Pierre Guerindon as he moved Caterpillar's manufacturing philosophy to the state-of-the-art level it enjoys today. The traits that had earned him national championship sailing titles in his native country of France were evident throughout his long, successful career at Caterpillar —but never more than during the 6-year period when he dramatically changed the company's basic approach to manufacturing.

Modernizing our global manufacturing base was not an easy task, but Mr. Guerindon met the challenge head on. Many in the middle and senior management ranks said it could not be done, it was sheer folly to attempt it, it was not cost-effective. But Pierre prevailed to his own credit and to the gain of the company.

Did we make mistakes? Did we spend more shareholders' money than we might have? Did the project take longer to execute than planned? The answer to these questions is yes. But the answer to the key question, *"Would we do it again if we had the option?"* is also an unqualified yes.

Pierre Guerindon was the architect and leader of Caterpillar's *Plant with a Future* and served his company extraordinarily well.

George A. Schaefer
Chairman of the Board, Retired
Caterpillar Inc.
Peoria, Illinois

Preface

When I left the Nippon Denso factory in Japan, all the pieces began to fall into place. Something big was about to happen to me—and to my company. I had just completed a two-year tour of about 100 of the world's most productive manufacturing companies: General Electric, Kobuta Motor, Kawasaki, Mitsubishi Motor, Toyota, General Motors, GM-Opel, Citroen, and Ferrari, among others. I had seen effective techniques in every location—robots, flexible manufacturing systems, automated material handling devices, advanced computer systems—but now I understood that it wasn't just technique that made these facilities efficient. The real keys to success were the *principles* upon which the plants were designed.

Nippon Denso had just displayed, with extreme simplicity, most of the fundamental principles of a continuous flow factory. Two principles were obvious: main production was flowing in a single direction and secondary operations were flowing into the mainstream harmoniously. I sensed the presence of other principles, but they were invisible. What I could see was the *effect* of these principles: simplicity, high quality, productivity with few people, zero inventory, high-speed manufacturing, reasonable automation, flexibility, fast cycle time. The plant operated like a ballet—a continuous and synchronous flow of parts, moving in one direction in perfect harmony, orchestrated by a continuous yet invisible flow of information. This was not a showcase of technology. In fact, there were no computers on the shop floor. Instead, there was a network of air-compressed piping with different-colored golf balls being blown through the system. Each color represented a different assembly, and each ball represented demand for 50 assemblies. When a ball was picked up at the piping terminal, a Kanban message was immediately put into effect, causing assembly to begin. As the assembly bins were depleted, balls of the same color were sent to internal and external suppliers, signaling them to produce and deliver the right type and quantity of replenishment parts. The entire chain of supply was synchronized by the message of the golf balls—the parts process, the rough store to feed the process, the purchase-finished store disbursement, and finally the suppliers. It was a superb application of Kanban with a high degree of velocity. Models changed several times a day, and the system was completely responsive to customer demand with no buffer inventory.

Some in our group thought the automation was too primitive and that computers could do a better job. Technically, they were right. But despite the unsophisticated environment, this was an impressive display of productivity. It convinced me that *true productivity results from a unique and fragile balance of operating principles.* I began to envision the day when my company would be guided by these principles. It would require a long journey, but the rewards would be worth the effort. Discovering and applying the principles of continuous flow manufacturing would give Caterpillar a major competitive advantage.

In the years that followed, my company and many others have attempted to achieve this advantage. Some have been successful. Others have not. Those that succeeded appear to have at least two things in common:

- A solid understanding of the fundamental principles of continuous flow manufacturing
- A corporate strategy and set of operating rules that creates an environment in which continuous flow manufacturing flourishes

This book addresses both areas. It is based on the ideas and experiences of approximately 2000 people working in more than 100 successful companies. One of those companies is Caterpillar Inc., the world's leading manufacturer of heavy equipment and a major supplier of diesel engines. Between 1986 and 1992, Caterpillar converted 18 factories in nine countries to continuous flow manufacturing. I am grateful to have served as the architect of this effort.

When I retired from Caterpillar, modernization was nearly complete. Quality had improved from already high levels. The new product development cycle had been dramatically reduced. The time required to build a product from rough material to shipping had also declined. Costs were trending toward the target, and inventory reduction promised to exceed our original goal. The company has been reorganized into profit centers that focus on return on assets. This reorganization has helped improve the effectiveness of our modernization plan. Most profit centers have completed their projects and achieved their goals.

In 1994, after achieving record-setting sales, profits and global market share gains in each quarter, Caterpillar Inc.

Chairman and CEO Donald V. Fites told financial analysts that the company's strategies are paying off. Fites said, "The company has improved its efficiencies over the years, primarily through the company's seven-year factory modernization program that today is showing a return of more than 20%. The modernization effort, of which the company continues to reap benefits, has improved Caterpillar's competitive position."

Although this book reflects many of Caterpillar's experiences, it is not a case study. It is a comprehensive review of the principles of continuous flow manufacturing and the management strategies that drive these principles. The book contains my personal vision, which should not be interpreted as that of Caterpillar management.

This vision is advanced, yet simple and logical. It has been implemented successfully in factories around the world, and it is as applicable in small businesses as it is in major manufacturing concerns. I present this vision with pride and humility, knowing that others may offer even better solutions. My hope is that this book contributes in some small way to a renaissance in American manufacturing.

I offer this modest contribution to my wife Arlette and the unique management team of Caterpillar Inc.

My sincere thanks and appreciation to those who contributed to this book:

- Managers from more than 100 continuous flow factories around the world who took time to disclose their principles and experiences
- My associates at Caterpillar Inc., who helped refine the vision
- Carol Leitch Fries, who helped restructure, edit, and prepare my manuscript for publication
- My editor, Russell Dekker, for his constructive criticism
- Machine, equipment, and systems designers who shared their knowledge and expertise, especially the following:

- *United States*: Ingersoll Milling Machines, Giddings & Lewis, Thermoprocess Systems (Holocroft), Ingersoll Engineers, McDonnell Douglas, Electronic Data Systems
- *United Kingdom*: British Aerospace, Ingersoll Engineers
- *Germany*: Scharmann GmbH, Fritz Werner, Huller Hille, Voest Alpine
- *Belgium*: Pegard
- *France*: Alsthom (C.J.P.), Citroen, Renault Machine Tool
- *Italy*: Mandelli Industriale Spa
- *Japan*: Mitsubishi Machine Tools, Mazack, Fanuc

Pierre C. Guerindon

Contents

Foreword   *iii*

Preface   *v*

Introduction   1

1. Strategies for a New Era   5

2. The Modernization Vision   17

3. The Continuous Flow Factory Vision   35

4. Automation for the Continuous Flow Factory   65

5. Overcoming the Challenges of Automation   103

6. Comparing Continuous Flow and Traditional
    Factories   127

7. Information Systems for the Continuous
    Flow Factory   141

8. Taking Action and Realizing the Gains   165

9. Consolidating the Gains of Continuous
    Flow Manufacturing   183

10. Justifying, Approving, and Monitoring
    the Investment   195

11. Creating a Continuous Flow Factory   217

    Conclusion   253

    Bibliography   *259*

    Glossary   *261*

    Index   *271*

Introduction

I. PURPOSE

This book was written to inform and educate readers about continuous flow manufacturing, a concept being applied by leading companies around the world. The message has been developed for manufacturing engineers, product designers, marketers, financial analysts, executives, corporate strategists, college students, and other interested readers. It presents a comprehensive discussion of the topic including:

- Key business initiatives that compete with and complement a continuous flow strategy

- The fundamental principles upon which a continuous flow factory is based
- The operating rules that create an environment in which a continuous flow factory will flourish
- Vital actions to be taken in preparation for modernization
- Technology and systems required to automate and integrate a continuous flow factory
- Processes for justifying the capital investment, approving modernization proposals, and monitoring implementation
- A methodology for modernizing an existing facility into a continuous flow factory

II. OVERVIEW OF BOOK

Chapter 1 sets the stage for modernization. It describes the changing competitive environment many companies face and outlines major offensive and defensive strategies a company might pursue. The chapter includes a discussion of how strategic options compete for resources and how they can be combined to leverage gains to the enterprise.

Chapter 2 provides an overview of the four key elements of the modernization vision: rationalize, simplify, automate, and integrate. It discusses the importance of preparing for modernization by rationalizing the product line and manufacturing capacity, and by simplifying products, processes, and procedures. The chapter also describes some of the fundamental challenges of automation and explains the role of the factory information system in continuous flow manufacturing.

Chapter 3 provides a detailed look at the principles upon which a continuous flow factory is based. It covers the basic principles of layout; flow; processing parts from start to finish; and pulling, rather than pushing, production. Also reviewed are principles related to material handling and scheduling, quality, suppliers, and human resources.

Chapter 4 is a technical discussion of automation. It describes manufacturing cells, transfer lines, cellular systems and flexible machining systems (FMSs), discussing each type of automation in terms of its strengths, limitations, and primary applications. The chapter also presents flexible welding and assembly concepts.

Chapter 5 explains how some of the unique challenges of automation can be overcome by using process innovations, such as flexible carburizers and head changers, and by applying critical layout engineering principles.

Chapter 6 provides a detailed comparison of a traditional and continuous flow factory, describing the key differences in automation, layout, operation, and management. Major advantages of the continuous flow approach are presented.

Chapter 7 covers the information systems that support a continuous flow factory. An overall framework for a realistic and achievable level of integration is presented, as well as a discussion of hardware, software, networks, and computing islands within a continuous flow factory.

Chapter 8 recaps all the fundamental principles of continuous flow manufacturing, plus the business operating rules which create the best environment for this type of operation. The chapter also translates the principles into actions to be taken and gains to be realized.

Chapter 9 provides objective data on the gains of a continuous flow factory. It explains the impact of modernization on a plant's total cost structure, outlines the gains that should be realized, and emphasizes how the gains can be leveraged when modernization is completed in conjunction with other strategic initiatives. Nonquantifiable benefits of modernization are also presented in this chapter.

Chapter 10 proposes a process for justifying the capital required to modernize a factory. It also provides a process for approving modernization proposals and includes detailed checklists to assist those who develop and review projects. A mechanism for monitoring the investment at the plant and corporate levels is also described.

Chapter 11 discusses a ten-step methodology for converting a traditional factory to continuous flow operation. It also highlights mistakes to avoid when developing and implementing a continuous flow manufacturing strategy and ways to prepare employees for the changes that will accompany modernization.

The conclusion presents a final message about continuous flow manufacturing from executives and consultants who have implemented this type of factory. It emphasizes critical success factors related to this strategy and speculates about continuous flow manufacturing in the twenty-first century. Also included are a bibliography, glossary, and index.

1

Strategies for a New Era

I. THE END OF AN ERA

The U.S. manufacturing industry enjoyed decades of uninterrupted growth and prosperity following World War II. But when the 1980s arrived, most manufacturers lost their sense of invulnerability. Almost overnight, demand for consumer and capital goods dropped as energy prices soared and major economies around the world slipped into recession. To make matters even worse, Japanese-based manufacturers charged into the U.S. marketplace backed by a weak currency, a strong government, and an intense commitment to exporting.

To remain competitive, long-time giants like IBM, General Motors, Caterpillar, and others downsized, reorganized, improved their products, and took major steps to cut costs. Many went to Japan to see their factories and study their processes. In the late 1980s, the world economy improved. Well-managed companies used that period of relative prosperity to modernize their manufacturing facilities.

Then came another deep, worldwide recession in the 1990s. Clearly, we have entered a new era. The good old days—those decade-long periods of nonstop growth—are over. Today, economic conditions fluctuate faster and more frequently. Companies are challenged to remain competitive throughout all the stages of the business cycle. This requires strategic planning.

Many companies have established a small group at the top of their organization to conduct corporate strategic planning. Strategic planning, with or without consulting help, is a fundamental aspect of managing the enterprise in today's new era. Selecting strategic initiatives, determining their sequence of introduction, assessing the combined effect of several options, and identifying the magnitude of change an organization can endure have become priority activities for top executives.

As a company's executives chart a course for the future, they assess a variety of strategic options. Those options fall into two basic categories—offensive and defensive. The distinction between offensive and defensive strategies is somewhat arbitrary, but in general, their differences are as follows:

Defensive strategies
- Offer quick fixes
- Result in change
- Emphasize cost reduction
- Minimize capital expenditures

Offensive strategies
- Offer long-term solutions
- Completely reshape a corporation

- Focus on changing the total cost structure
- Require investments for the future

Although there are important differences between offensive and defensive strategies, companies can (and should) pursue both.

The primary focus of this book is an offensive strategy—plant modernization through implementation of a continuous flow factory. Other offensive and defensive options are reviewed because:

- Investing in a continuous flow factory competes with other initiatives for the organization's limited financial and human resources.
- Before approving capital expenditures for a continuous flow factory, top management will want to review alternative strategies.
- By combining two or more strategic options, the company may leverage the benefits of both.

II. DEFENSIVE STRATEGIES

During the deep worldwide recession of the 1980s, sales and profits declined dramatically for many major corporations. Many companies incurred losses and were forced to launch aggressive cost reduction campaigns to stop the flow of red ink. They consolidated operations, contracted work out to lower cost producers, closed plants, reduced inventory, and slashed capital expenditures. These actions helped many companies survive. Organizations operating in the survival mode typically pursue defensive strategies. We will briefly review five basic defensive strategies and the impact each has on modernization.

A. Merger, Acquisition, Strategic Alliance

This option involves joining forces in some manner with another organization. Companies generally pursue a partnership in order to gain relatively fast access to new products, markets, or technologies. Organizations also establish alliances to acquire additional resources, strengthen their balance sheets, or profit from the reselling of portions of an acquired company. If management is pursuing an alliance with another organization, the odds are high that no major modernization project is being considered. However, following a merger, acquisition, or strategic alliance, a company may choose to modernize. Before modernization can take place, management must review the product lines, supplier networks, manufacturing facilities, and distribution channels for possible rationalization.

B. Emigration

Emigration involves moving production to a new location. In recent years, the trend has been to move to areas with lower cost labor, either outside the United States or to the Southern or Southeastern United States. Very few large corporations have relocated their entire manufacturing base. Rather, they move certain labor intensive portions to lower cost regions. For example, electronic components manufacturing has shifted to Mexico and the Far East. Fabrication operations have been relocated to Central and Eastern Europe or Mexico. North Carolina has become a popular area for basic manufacturing.

The primary goal of an emigration strategy is reducing labor costs. Some companies report realizing a 40 to 50% reduction. However, there are trade-offs associated with moving manufacturing away from product engineering, corporate headquarters, suppliers, distributors, and customers. Communication can be difficult. Logistics costs may increase. Responsiveness to customers may decline. These problems and others have the potential to partially offset the labor cost reduction. Despite the trade-offs, emigration can be an effective strategic option. It may be used in conjunction with a modernization strategy, emigrating one portion of the manufacturing base, while modernizing another.

C. Continuous Cost Reduction

Continuous cost reduction is an annual incremental gain in productivity. In the 1980s, every corporation had the potential to reduce costs and eliminate waste. Many still do today. Key targets for reduction include the cost of scrap, rework, and warranty; labor costs associated with excess employment; and materials costs.

Continuous cost reduction is not a natural behavior because most managers believe they operate at the right level of employment and manage their costs carefully. Therefore, to achieve significant reduction it is usually necessary to mandate cost cutting from the top of the organization. A mandate may take many forms, including:

- A directive to cut the budget by 5% or more
- Incentives for early retirement
- Layoffs and/or plant closings
- Wage and salary freezes or cuts
- A moratorium on hiring
- Reductions in travel
- Programs aimed at reducing suppliers' costs and prices
- Reductions in capital expenditures

In the past ten years, most major companies have used one or more of these tactics. They have been effective, especially when a company's costs and prices are not competitive and when financial results are poor. If an organization seriously engages in ongoing cost reduction activities for three to four years, a cost-conscious environment can be created, and cost reduction can become ingrained in the culture. This can produce a total reduction of 15 to 20% or more. This reduction is actually a cost squeeze, not a permanent change in cost structure. But the benefits flow directly to the bottom line and no investment is necessary.

A continuous cost reduction strategy may appear to conflict with a modernization strategy because modernization

often increases costs in the short term. However, modernizing has the potential to reduce costs in the long term, and
can therefore enhance continuous cost reduction efforts.

D. Process Simplification

After several years of aggressive cost cutting, most corporations find there is limit to the gains that can be realized. To
achieve further reductions, they must pursue more creative
approaches such as process simplification.

In most corporations, business processes have evolved
over many years and have become complex and complicated.
Elaborate systems of checks and balances have been put into
place to assure that all transactions are completed with
integrity. While these procedures and systems have been
effective—helping companies operate ethically, avoid mistakes, and maintain internal control—they have not always
been efficient. In many cases, complex processes have only
served to generate a high-cost, bureaucratic paper mill.
Paper processing is slow in most of these mills, especially at
high levels of the organization.

To speed transactions and improve quality and efficiency, processes must be simplified in all areas of the company. For example, the typical company's process for paying
supplier invoices is often highly functional, involving
employees from accounts payable, purchasing, and quality.
This process can be streamlined by locating all functions in a
single cell (Figure 1). A cell team, responsible for a certain
quantity of part numbers, handles the entire accounts
payable process from start to finish. Computer terminals in
the cell give team members access to functional information.
By sharing information and communicating face-to-face—
rather than through memos and phone calls—problems can
be resolved up to ten times faster than with the previous
process. As a result, the number of employees required to pay
supplier invoices can be reduced by 60% or more, and those
who remain have higher job satisfaction because they are
accountable for an entire process—not just a piece of it.

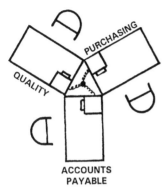

Figure 1 Office cell.

There are many other opportunities for process improvement in the office. Research indicates that a medium-sized production plant has more than 100 different technical, financial, engineering, logistics, and personnel processes, most of which are candidates for simplification. The key to simplification is employee involvement. Empowering employees to simplify their jobs results in more efficient and cost effective processes. It also improves morale and allows change to be implemented faster because employees have ownership of the new processes. There is a cardinal rule of work simplification: reward innovators. Don't lay them off. If people perceive that their jobs are in jeopardy, they have no reason to participate in work simplification. Process simplification can be a low-cost, high-return method of achieving cost reduction. It complements—but does not compete with—a modernization strategy.

E. Diversification

Diversification involves investing in a different type of business. The goal is to enhance corporate profitability and generate a better return on assets by stabilizing revenues throughout the entire business cycle. Although diversification was popular during the 1980s, there are few success stories. There is growing consensus that it is difficult to be profitable in a field in which you have no experience. Many corporations that once diversified are now returning to their core businesses and pursuing other strategies for leadership. Diversification requires a commitment of financial resources and therefore competes with a modernization strategy.

III. OFFENSIVE STRATEGIES

After implementing aggressive cost reduction programs, some companies emerged from the recession of the 1980s with the financial strength to develop and pursue offensive business strategies. As stated earlier, offensive strategies are longer-term initiatives that reshape a corporation and permanently change its cost structure.

A. Reorganization

Most large companies have reorganized at least once during the last ten years. Some, like IBM, continue to change their organizational designs to improve performance. Companies that reorganize have several objectives, including:

- Eliminating bureaucracy
- Reducing employment levels
- Increasing management span of control
- Pushing decision-making to lower levels of the organization
- Shortening new product development cycles
- Improving responsiveness to customers
- Improving profitability

One of the most common methods of reorganization is dividing a company into profit centers or business units. Each unit is accountable for developing and marketing a specific product line in a manner that satisfies customers and earns an acceptable profit. Business units are multifunctional work groups with representatives from many disciplines. Human resources are concentrated in these units with corporate staff reduced to a minimum. Only functions that can be performed economically for the entire enterprise, such as Legal, Treasury, and Public Affairs, are retained at the corporate level.

Each business unit sets its own financial goals and develops strategies to achieve them. Top management reviews business unit goals and strategies regularly to assure they support enterprise objectives. Incentive compensation is used to foster an entrepreneurial environment in the business unit. Some companies link as much as 20 to 50% of total compensation to the achievement of approved financial goals. Profit centers are most effective when there is very little exchange of components between business units. When a company is highly integrated, and a large number of components are transferred between business units, internal transfer prices can distort profit levels. The process of negotiating internal transfer prices can cause conflict, as the selling unit strives to increase revenues and the purchasing unit focuses on lowering costs. Conflict between business units can be positive, forcing cost reduction of internally sourced components. But conflict can also cause problems, including:

- Too much emphasis on numbers. When business unit managers make decisions solely on the basis of market price, they may fail to consider the additional value included in internally made components. Engineering and manufacturing innovations can enhance value and command a higher price, as long as the value is perceived and wanted by the customer.
- Too much focus on the short term. Profit center managers may become driven by short-term results, particularly when their incentive compensation checks are at risk.

To avoid these conflicts:

- Top corporate management must keep a long-term perspective, reviewing profit center goals and objectives regularly to assure they support the enterprise.

- People selected to run the business units must demonstrate an ability to work effectively in the short term, while maintaining a long-term focus.
- A portion of each profit center's incentive compensation package can be linked to enterprise profitability, reinforcing the need for each business unit to support a long-term corporate strategy.

B. Modernization

Companies define modernization in many ways. Some modernize by purchasing machine tools. Some set up manufacturing cells. Others invest in multimillion dollar Flexible Machining Systems (FMSs). Generally, most companies identify modernization with hard automation and large capital expenditures. However, investment in machinery is only one element of an effective modernization strategy. New machinery can create process flexibility, but it only affects direct labor costs, which amount to just 10 to 15% of total costs. Much greater gains can be realized through a broader approach to modernization which includes:

- Factory layout
- Logistics systems
- Supplier programs
- Design simplification
- Product standardization
- Human resource utilization
- Quality management
- Customer responsiveness

A modernization strategy that includes all these elements can result in a cost reduction of 25% or more. And if such a strategy is combined with defensive cost cutting options, the potential gain can be as high as 40%.

Despite these benefits, there is pain associated with implementing a broad-based modernization strategy. The remainder of this book is devoted to reviewing the challenges and rewards of factory modernization.

IV. SUMMARY

When contemplating a modernization strategy, a company should review the plan in context with other major strategic options.

Strategic option	Impact on modernization
Merger	Affects modernization if new partner has a similar product line. Prior to modernization, companies must rationalize products, facilities, suppliers, organizations, and distribution channels.
Emigration	May be feasible to emigrate a portion of production and modernize remaining manufacturing base. Must consider additional logistics costs and ability to be responsive to customers.
Continuous cost reduction	Highly desirable program that can stand alone or be combined with modernization.
Process simplification	Critical, ongoing initiative that stands alone or enhances modernization efforts.
Diversification	Has no bearing on modernization other than to compete for the corporation's limited resources.
Reorganization	Can achieve excellent results, especially when implemented in conjunction with a modernization strategy.

The ultimate cost-killer is a combination of four options:

Option	Potential cost reduction
Modernization	25%
Continuous cost reduction	5% per year
Process simplification	5% per year
Emigration	Depends on circumstances

V. BUSINESS OPERATING RULES

R1. Business operating rules define an environment in which a continuous flow factory will achieve maximum gains.

R2. Modernization should be considered in context with other strategic options.

R3. The benefits of modernization can be leveraged by combining it with other strategies such as continuous cost reduction, simplification, emigration or reorganization into profit centers.

2

The Modernization Vision

I. ELEMENTS OF MODERNIZATION

The modernization vision presented in this book has four elements, as shown in Figure 2. They are:

- Rationalize
- Simplify
- Automate
- Integrate

Although these four words sound simple, they represent a comprehensive vision of a productive and profitable manufacturing operation. This operation has the following characteristics:

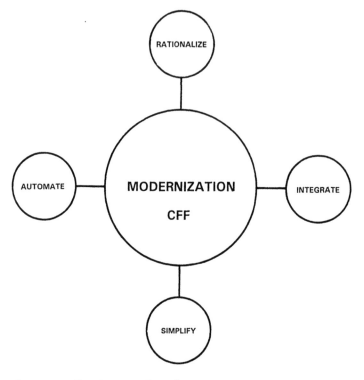

Figure 2   Continuous flow factory concept.

- Its capacity is sized appropriately.
- The products it makes reflect customer needs.
- Only the "vital few" components are being made in-house.
- It has built strong relationships with a small number of suppliers.
- Products are designed for manufacturability.
- Business processes have been streamlined.
- Factory layouts are efficient.
- The appropriate level of automation has been purchased and installed.

- An electronic network links the company's factory floor with engineering data and internal and external logistics systems.

II. PRELUDE TO MODERNIZATION

The modernization vision described in this book involves far more than automation. Automation is vital part of the vision, but it is not the focus. Automation cannot solve every problem in a factory. If a facility has a bad layout—if operations are located far apart and in the wrong places—automated material handling will not correct the problem. The real solution is to rearrange the plant layout to avoid material handling and minimize the need for automated transportation. Likewise in the office, automation is not always the answer. It does not pay to computerize a labor intensive clerical function without first setting better work rules.

To make automation work—and make it pay—a company must first rationalize and simplify the operations which will be modernized. Ideally, rationalization and simplification are the prelude to automation. They happen first before any purchases are made. But in the real world there is pressure to automate quickly, so many companies do not take time to complete the prelude before investing in new machines and systems.

Despite the pressure to automate at a rapid pace, modernization will pay greater returns if the organization makes a concerted effort to rationalize and simplify before automating. These efforts will actually save time in the long run and reduce the total cost of implementation.

III. RATIONALIZATION

Rationalization begins with the marketing strategy, and carries through to the product design and production strategies.

A. Marketing

Before embarking on a modernization plan, the organization must rationalize its product offerings. This requires developing a strategy that outlines which market or markets the company will serve and which product or products it will offer. There are many ways to set this strategy, but the basic process involves five steps:

- Divide the total market into segments, grouping customers by industry, product application, geography, or other common factors.
- Determine customers' current and future product and service needs for each segment.
- Analyze competitors to determine the company's relative strengths and weaknesses.
- Assess sales and profit opportunities in each segment.
- Select segments to serve and products to offer based on competitive strengths and profit potential.

B. Product Design

Once the marketing analysis has been completed, the company can establish a product strategy for the future. Winning products will be defended, and losers either fixed or dropped from the line. A new rationalized product line emerges, enabling the firm to focus its limited design resources on enhancing leading products and improving losers that exhibit long-term potential.

C. Production Planning

After identifying a new rationalized product line for the future, the company must size manufacturing capacity appropriately. Many manufacturers—particularly large ones —have too much capacity. They made large increases during the 1960s and 1970s, and although they downsized in the 1980s, many did not go far enough. Today their markets are smaller, there is more competition, and growth is slower than anticipated.

Excess capacity absorbs capital and drains profitability through the costs associated with low utilization of facilities. Very little can be done about the cost of low utilization, apart from eliminating variable labor, which is helpful but generally not sufficient. Excess capacity can often be attributed to overly aggressive marketing forecasts. When considering long-term capacity requirements, it is important to trim marketing forecasts. The best course is to size capacity conservatively, leaving some unallocated space for small, future adjustments. A conservative capacity plan may cause the company to lose sales during peak demand years. But in most cases, the benefits of a more cost-effective long-term capacity plan outweigh the costs of business lost in peak years. In addition, the few customers lost during a boom year may be persuaded to come back when product availability improves.

Once the product strategy and capacity plan have been determined, the next step is to size the manufacturing base appropriately. This may require closing or consolidating facilities. Closing a plant is one of the most difficult and painful actions a company ever takes. The negative effects include:

- The impact on employees, families, and plant communities
- The short-term reduction in profit related to the write-off of the facility
- The costs associated with moving products to other facilities

On the positive side:

- Period (fixed) labor costs decline as management, technical, clerical, and indirect hourly employment rolls are reduced.
- Period expenditures like depreciation, utilities, maintenance, and subcontracted services are also reduced.
- Capital is no longer required to maintain the facility.

- Cash may be generated through the disposal of surplus production assets and property.

Overall, a plant closing can be a very efficient way to reduce costs. To assure maximum cost efficiency, the facility receiving the product from the closed plant, if underutilized, should commit to transferring or hiring only direct labor for production. This will have a dramatic effect on total plant efficiency by better utilization of all existing services. People, machinery, equipment, computer systems, and other assets can be spread over a larger base, lowering the cost of the product being transferred as well as all other products made at the facility. Additional cost savings may also be realized as a plant prepares for incoming product. Experience shows that when plants get ready to receive new products, other problems are often identified and resolved, generating unexpected cost savings. There is no debating the fact that plant closings are painful, but in the final analysis, they are often a vital step in a comprehensive modernization strategy.

IV. SIMPLIFICATION

Simplification is another vital aspect of the prelude to modernization. Ideally, all processes should be simplified before they are mechanized. But the reality is, most companies automate before they are finished simplifying. This is not a major problem as long as they do two things:

- Prior to modernization, simplify those things that have a major impact on the program.
- Continue to simplify after modernization.

This discussion will focus only on things should be simplified during the prelude to modernization.

A. Make/Buy Decisions

Make/buy decisions are a critical aspect of simplification. Historically, companies have made too much in-house, thinking that volume was "good for the numbers." (Perhaps they were looking at the wrong numbers!) The determining criterion should be cost, not ratios that measure shop floor activity as a form of productivity. There is great simplification in making only the "vital few" parts the company has to make because it cannot buy them. This is much more efficient than the past policy to buy only what could not be made.

For best results, make/buy decisions should be made by sourcing teams made up of representatives from several disciplines. Engineering can provide input on designs that are proprietary and should be made internally. Manufacturing has information on process development. Purchasing contributes supplier data. Accounting maintains the cost perspective. Together these multifunctional sourcing teams can evaluate families of parts and components and ultimately determine a "core make" list and a "core buy" list. Future decisions can be extrapolated from these lists. Using this process, a company can reduce the number of parts it makes by 50% or more. The goal should be to produce in-house only those unique parts that cannot be purchased. This is typically 20% of total part numbers and represents 80% of total costs. By focusing manufacturing resources on the "vital few," modernization will be faster, easier, and more cost effective. Making only the "vital few" parts also helps the organization maintain a conservative capacity plan and operate with less manufacturing space.

Once a make or buy decision is reached, it is important to retain stability. This can be a problem, especially if a supplier prices its parts low to get the company's business, then raises prices when the company has disposed of its internal processes. Generally, it is safer to base a make/buy decision on internal accounting numbers, and at the same time, identify the root cause of the supplier's better cost. One word of caution: although the internal accounting group has an important role to play in a make/buy decision, the final decision should not be made by numbers alone. A team approach typically produces better results. To assure long-term stabil-

ity of make/buy decisions, it is critical that product and process engineers work together to continuously improve the
manufacturability of all parts—those produced internally as
well those made by suppliers.

Assuring price stability can be even more challenging
when suppliers are located outside the United States. It is
not practical for companies to reverse sourcing decisions
when exchange rates fluctuate, because by the time the
organization tools up to produce the part internally,
exchange rates may have changed again. To maintain stability, make/buy decisions should be based on equilibrium
exchange rates. Equilibrium rates are long-term rates that
consider the effects of inflation in a country compared to
inflation in the United States. If a company is currently producing a part in-house, but exchange rates make outsourcing
attractive, it may choose to slow down or mothball production and buy temporarily from the outside. Keeping the
assets in place helps deter supplier price increases and
hedges longer term costs.

B. Supplier Consolidation

Although simplification leads to more purchases from suppliers, it does not increase the size of the supplier base. To the
contrary, the number of suppliers should be reduced. Many
companies have too many suppliers. Over time, they have
added new sources without abandoning old ones. And they
tend to keep their options open, anticipating exchange rate
fluctuations, changes in performance, or supplier financial
difficulties.

By keeping a maximum of two or three suppliers per
commodity, the company gets more purchasing power. It also
retains the high volume needed for convincing suppliers to
endure the pain of just-in-time delivery, packaging in known
quantities, bar code identification, and quality certification—
programs essential to an effective modernization plan. With
fewer suppliers, the company is better able to enter into
partnership arrangements like those found in Japan. In

these special relationships, suppliers become an extension of the corporation and behave accordingly. Cost information is shared, designs are developed cooperatively, and suppliers do not compete in the after-market parts business. Partnerships such as these are based on mutual respect and ultimately serve the best interests of all parties. Consolidating the supplier base is critical to developing these kinds of partnerships. Defining a core group of committed supplier partners is a fundamental aspect of simplification, and a vital prelude to modernization.

C. Simultaneous Engineering

Simultaneous engineering means that products and manufacturing processes are designed at the same time, by the same group of people. Traditionally, most companies have used sequential engineering. With this approach, product design comes first and is done by the engineering function. Process design follows, and is the responsibility of the manufacturing discipline.

Since the mid-1980s, many companies have attempted to do more simultaneous engineering. They have created teams of product design and manufacturing engineers and charged them with the responsibility of developing products that meet customers' expectations and can be manufactured cost effectively. Some of these teams have been successful, but others have struggled. Those that have struggled have two things in common:

- Manufacturing people do not have the background to communicate effectively with engineers.
- Manufacturing people wait too long to select a process. Their input is critical during the concept stage. If they wait until a later stage, engineering is often reluctant to accept changes.

Simultaneous engineering is being used most effectively in European and Japanese companies where most manufacturing planners have engineering degrees. It is also successful in organizations where production areas are handled by people with engineering backgrounds. The number one factor contributing to the success of simultaneous engineering teams is early involvement of manufacturing engineers who understand the engineering discipline. Supplier engineering involvement is also valuable.

Simultaneous engineering is a key element of a total modernization strategy. If a company is purchasing new machine tools and systems, it is vital that products be designed to take advantage of new process capabilities. Basic guidelines for simplification include:

- Reduce part numbers.
- Standardize material.
- Minimize specifications for forging and casting.
- Standardize plate chemical specifications and thicknesses.
- Assure that the machinery for the process and the tolerance for the design are in harmony.
- Use design to prevent assembly errors (e.g., use asymmetrical design to eliminate confusion about up and down positioning; locate holes appropriately; design for robotic assembly).
- Design for supplier manufacturability.

Although simultaneous engineering is an important part of the prelude to automation, it is an excellent discipline to introduce even if there are no plans for modernization. The same is true for other simplification activities and for the rationalization effort. Any company can benefit by rationalizing its product line, manufacturing capacity, and facilities, and by reducing the size of its "core make" list and its supplier base.

V. AUTOMATION

Automation is typically the most visible aspect of a modernization plan. It is also the most controversial. In recent years, the business press and investment community have criticized American manufacturers for allocating too many resources to factory floor operations. Major concerns about automation include the high cost of capital, costly problems associated with start-up, the negative impact on short-term profitability, and the failure to achieve projected results. Some of these criticisms are valid. Automation is costly. It does erode short-term profits. And in some cases, companies do not achieve the results they had anticipated. But when automation is viewed as a medium- to long-term effort—a five- to six-year commitment—and when it is executed properly, it will position the company for profitable long-term growth.

The largest single factor affecting the success of automation is the attitude of those purchasing new technology. In many companies, there is a desire to be first—to use leading-edge technology, to develop a "Star Wars" factory. These companies lose sight of their original objective—cutting costs—and focus instead on creating an impressive machine show. They push their vendors beyond their capabilities and the end result is often higher costs, lengthy delays, longer start-up, and reduced benefits. Successful automation is reasonable. It is state-of-the-art, but not futuristic. It avoids gadgets and prototypes, concentrating on proven technology. Prototypes have their place in manufacturing research and development, not on the shop floor.

In Japan, manufacturers tend to recognize the value of proven technology. One company installed and ramped-up production on a five-machine flexible machining system (FMS) in three weeks. That process can take up to three years in some U.S. companies. The difference was, in Japan, the FMS was standard and had been proven in other applications. Most U.S. manufacturers avoid standard FMSs, preferring full prototypes.

Some may argue that there are many cases where standard technology is not acceptable. This argument is valid, but the point is the number of changes should be limited. When companies avoid risky applications of technology and use proven solutions with no more than 20% of the content changed, installation and ramp up time are reduced, machine uptime increases, and bottom-line results improve.

Automation has received some bad publicity in the 1980s and 1990s. But automation is not the problem. People are the problem. The solution is using proven technology. It is easy to lose perspective and be overly impressed with the latest new probe or "fuzzy logic" software. But the fact is, machinery is merely a necessary evil. It helps create the flexibility and flow needed for optimum efficiency. Machinery alone does not produce much cost savings. Savings are realized mostly through the combination of layout, machines, processes, logistics, and systems.

Good automation is the deployment of machinery, processes, and material handling, all coordinated by computer. Production materials flow efficiently from the point of reception to the point of use. And machinery is laid out in such a way that a family of parts flows continuously from rough to finish without stopping. The parts families merge with others and with purchased finished goods, ultimately feeding the assembly line just in time. Automation will be discussed in more detail in Chapters 4 and 5.

VI. INTEGRATION

The final dimension of modernization is integration. Integrated systems create a flow of information that parallels the flow of material. The information flow is invisible, but it causes the material flow. Integrated systems control external logistics, coordinating delivery from suppliers who feed the plant with rough or finished parts as needed. Integrated systems also drive internal logistics, regulating material movement in the plant to feed the assembly line. They link stores and processes with material handling devices, creating a continuous and synchronous flow with minimum inventory.

Another role of integrated systems is computer program monitoring. An integrated factory information system supports computers on the shop floor, downloading machine programs, changing tooling, and providing gauges and instructions to operators.

Integration links machines on the factory floor with internal and external logistics systems and automated material handling equipment. This allows material to move through the plant with very little human intervention. Humans supervise and monitor the flow of material, but do not cause it.

Because material flow is regulated by computer, many jobs are no longer needed. This represents a major cost savings. However, in order to achieve the full savings, all aspects of the factory must be integrated. If any elements are excluded—internal or external logistics, material handling, or computer program monitoring— manual intervention will be required, diminishing the return on the investment. Automated material handling is one aspect of factory integration that is frequently cut. When a project exceeds the budget, there is pressure to eliminate some portion of the investment. Automated material handling is often the prime candidate. Also, automated material handling is normally the last aspect of the job to be completed, making it a target for delay or cancellation.

When a company invests in new machinery, but does not automate the material handling, it will not achieve the expected results. Machinery requires about 80% of the total investment and generates approximately 60% of the benefit. Only by making the additional 20% investment in material handling can the company realize the full benefit of modernization. When budget pressures arise, it is therefore more cost effective to buy fewer machines and complete their installation with logistics and material handling. By keeping old machines longer and redeploying them with new machines in a more efficient layout, the company will get the full benefits of employment and inventory reduction, higher velocity work flow, and quality improvement. Integration will be covered in greater detail in Chapter 7.

VII. THE MULTIPLIER EFFECT

The four dimensions of modernization—rationalize, simplify, automate, and integrate—have a unique multiplier effect. Completing the first two dimensions in advance multiplies the benefits of the last two dimensions. For example:

- If design standards are simplified, automation is better.
- If the number of suppliers is reduced, integration of external logistics improves.
- If fewer parts are made and capacity is rationalized, capital requirements are reduced and automation is more efficient and cost effective.

There are many other examples. All confirm that it pays to exercise patience in the prelude phase (rationalize and simplify). Direct entry into the second phase (automate and integrate) has the potential to waste capital. All four dimensions of modernization are critical, but without the prelude, full benefits can not be achieved.

VIII. SCOPE, SEQUENCE, AND TIMING OF MODERNIZATION

There is no single best way to determine the optimum scope, sequence, and timing of modernization. At the minimum, the scope should include a product and the facility that makes it. Because modernization requires changes in layout, business processes, suppliers, material handling, computer systems, and other areas, the scope must include an entire facility— not just part of one. The size of the facility being modernized is not important. Scope is what provides the competitive edge.

Scope is particularly important in a highly integrated company where plants produce components for other plants. For example, an auto company that modernizes a transmission plant is wasting resources if engine and assembly plants are costly and out of date. In these situations, the scope of modernization must be larger. As for the sequence, the ideal approach is:

- Set marketing and product strategies.
- Size capacity and manufacturing base.
- Design products and processes.
- Modernize facility.
- Allow autonomy for supplier development.
- Allow autonomy for systems integration.

Determining the appropriate timing for modernization is most challenging when there is more than one plant involved. Companies may choose a step-by-step approach, modernizing one plant at a time. Or they may take the "big bang" approach and modernize all facilities at the same time. Some of the conditions that favor the incremental or step-by-step approach include the following:

- The business is generally successful. Cost and quality are out of line in selected areas only, so the company can modernize in small segments of the business.
- The company is not highly integrated. Few parts are transferred between units. The costs in one unit do not affect the costs in others.
- There is no emergency. The company's goal is to preserve leadership and it can select specific areas for quantum improvements.
- Capital is scarce and the balance sheet requires attention. Debt is not welcome.
- The prelude to automation (rationalizing product line, reducing capacity, eliminating manufacturing space, simplifying design, consolidating suppliers, etc.) will provide sufficient cost reduction, preventing the need for automation and integration.
- The company is organized in business units and each unit manager has the authority, vision, and experience to modernize as needed.

Conditions favoring the big bang effort:

- The business is in jeopardy. The company has lost leadership in spite of other cost reduction efforts. It needs a bigger productivity gain than can be achieved through incremental modernization.
- The company is highly integrated. Cost reduction and quality improvement are needed on all aspects of the product.
- The company has cut prices faster than it has cut costs and needs a quantum improvement in cost to survive.
- The company has exhausted other cost reduction efforts and needs further gains.
- Capital is available for investments promising good returns.

Obviously, there are no easy answers to the question of timing. Many factors must be considered and, in the final analysis, top management must make a judgment call. One thing to keep in mind: the incremental approach could take up to 10 years to complete. During that time, economic conditions will have changed—probably more than once. In addition, the program could lose momentum. Enthusiasm could erode, and the vision could be lost, particularly if the champions of the project have moved on to other jobs. When a project continues for many years, the risk is high that some aspects such as material handling and integration, will be dropped —diminishing the overall return on the investment. For these reasons, it may be more effective to pursue the big bang approach if capital is available.

IX. SUMMARY

Key points of the chapter:

- The modernization vision has four dimensions. Rationalize and simplify, the first two dimensions, constitute the prelude to modernization. Automate and integrate are the second phase of the vision and represent the essence of modernization.

- The prelude involves rationalizing the product line, plant capacity, and manufacturing space. It also includes simplification efforts such as reducing the number of parts made internally, consolidating the supplier network, and engaging in simultaneous product and process engineering. These activities can produce impressive gains in productivity, even when no modernization is planned.
- The second phase of modernization begins with investing in new machinery and systems to automate production and material handling. The material handling portion of the investment is vital to achieving maximum returns. The final step in modernization is integrating machines, material handling, and logistics systems with a factory information system. Integration is difficult, but is also key to maximizing returns.

X. BUSINESS OPERATING RULES

R4. A continuous flow factory is designed after conducting a comprehensive market survey, developing a product strategy, and defining a product line for the future.

R5. A continuous flow factory operates with good capacity utilization, but is capable of growing by steps within unallocated space.

R6. A continuous flow factory makes a minimum number of parts based on economies of scale and proprietary design.

R7. A continuous flow factory is very closely tied with a small number of competitive suppliers who enjoy a volume advantage and commit to vital practices such as just-in-time delivery, quality and quantity certification, and electronic identification of parts.

R8. A continuous flow factory uses simultaneous product and process engineering to design for manufacturability. A design for supplier manufacturability program is also in place.

XI. FUNDAMENTAL PRINCIPLES

F1. Select a reasonable level of automation using proven technology.

F2. Use standard machines when possible and avoid pushing vendors beyond their experience and capabilities.

F3. An integrated factory information system is needed to coordinate delivery from suppliers, regulate movement of material, and support the processes by machine program download and tool management.

F4. Regulate the continuous flow factory by computer. Human intervention is necessary to supervise and assist flow, not cause it.

The Continuous Flow Factory Vision

I. THE PRINCIPLES OF CONTINUOUS FLOW MANUFACTURING

The modernization vision described in Chapter 2 presents a broad picture of a manufacturing operation that has been rationalized, simplified, automated, and integrated. This chapter provides a more detailed look at that operation, outlining the major principles upon which a continuous flow factory is built. The principles presented here were acquired by visiting more than 100 manufacturing facilities around the world. No single plant is currently practicing all of these principles. Each operation has made compromises in some areas and therefore falls short of the total vision. However, the vision can be viewed in its entirety by looking at a collection of the best practices from the world's best manufacturers.

These practices were not immediately obvious to us during our plant visits. At the time, we were focused on automation and were not perceptive enough to see the revolutionary concepts underlying the operations. But after many visits and much analysis, the truth began to reveal itself. We realized that flexible machines and automated equipment are just one element of an efficient and profitable manufacturing operation. Equally important are the topics covered in this chapter: layout, flow, computer-controlled material handling, integration of logistics, systems that regulate flow, process control, supplier relationships, employee motivation, and more.

II. BASIC PRINCIPLES

A. Layout

A good layout is the most important element of a continuous flow factory. Most of the older plants we visited, particularly in the automotive industry, had poor layout. The reasons were historical. Over the years, models and processes have changed frequently. There was never enough time for rearrangement, so new processes were located wherever there was room—not necessarily in the most efficient locations. The factories we toured had compensated for their layout problems with miles of conveyors, so work flow appeared to be efficient. But automated material handling equipment is an expensive solution to the problem of poor layout. To avoid the cost of conveyors, *rearrange the plant into a continuous flow layout. Work should flow in one direction, and processes should be located in such a manner that material handling is minimized or avoided.*

B. Pull of Flow

Work flow in a continuous flow factory can be compared to a river. River water is not pushed into the ocean, it is pulled. When river water enters the ocean, gravity pulls more water from secondary sources and replenishes the river. This process is continuous.

In a continuous flow factory, the assembly area is like the ocean. As parts are consumed, assembly must be replenished, so internal and external suppliers are signaled to produce new parts. These parts are pulled into the flow and they move to assembly where they are consumed, once again setting off the signal to suppliers to replenish the flow.

Figure 3 illustrates a factory utilizing the pull concept. The action starts at the bottom of the layout on the main assembly line. The assembly program includes six days of orders. Subassemblies are fed to assembly half a day ahead of need. As assembly is performed, parts are consumed from the assembly sequencer nearby. Parts are also consumed from the sequencer as subassemblies are produced. A few large parts do not enter the sequencer and are routed directly to the main assembly line. The assembly sequencer is fed in two ways. Purchased finished parts are delivered just-in-time from suppliers to the docks on the right. They travel by conveyor to the sequencer. Worked parts are delivered from the three minifactories at the center of the layout. Machining and heat treat are done in minifactories 1, 2, and 3. The fourth minifactory does tacking, welding, and machining by two flexible machining systems. To feed the minifactories, there is a rough material sequencer near the top of the layout. The rough material sequencer is fed from the press shop and from four docks where rough materials have been delivered just-in-time from suppliers. Sequencers are well placed to minimize material handling as the flow moves from top to bottom of the layout. In this example, the rough material sequencer also performs the role of welding sequencer, disbursing parts as needed from the press operation to the tacking and welding station. Parts flow through

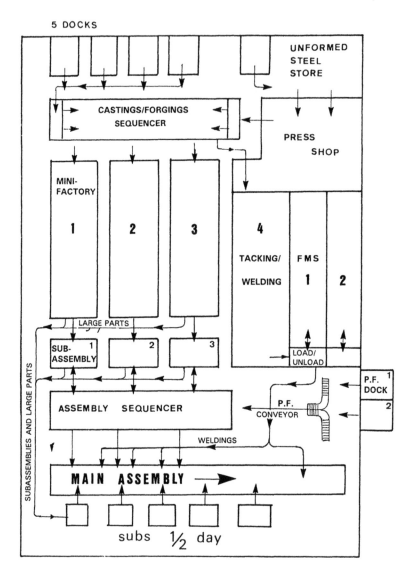

Figure 3 Continuous flow factory block layout.

this operation in a smooth and uninterrupted manner, responding to the needs of assembly.

C. Maintaining Flow

Water in a river moves naturally in one direction, following the law of gravity. Likewise, in a continuous flow factory, parts are processed in a natural way. They present themselves and are completed in one continuous process, in the order of their arrival and position on the conveyor.

In a continuous flow factory, managers do not schedule the iron and create the flow. *Their job is to maintain the flow.* There are many forces that will stop the flow, such as quality problems, machinery failures, and human errors. It is management's job to remove those obstacles and keep the flow moving.

D. Elimination of Inventory Buffers

The "river" that runs through a continuous flow factory maintains a low water level. There are no large inventory buffers, only a small safety stock to support two to three days of production. Inventory buffers have been eliminated because they are expensive and they conceal problems. As the management of Toyota in Japan told us: "When the river is low, you see the rocks and correct the situation. But if the water level is high, the rocks are never in sight, so you live with your problems, rather than fixing them."

E. Processing Parts

Water in the river always flows from a starting point to an ending point. It does not stop and rest. Likewise in the factory, work flow must allow parts to be processed from start to finish with no stops. This means all operations—including welding, machining, heat treating, and chemical processing (like chrome plating)—must be integrated into the overall process. This is a big departure from the fragmented, stop-

and-go flow being used in many factory operations. In these operations, parts undergo one process at a time and are stored, at least temporarily, between processes. This increases manufacturing time, as well as storage and material handling costs. It also contributes to scheduling conflicts and quality problems.

F. The Right Machinery

A detailed discussion of automation for a continuous flow factory is presented in Chapter 4. In general, the machinery used in a continuous flow factory must have several characteristics:

- *The process should be flexible* and capable of handling a variety of parts in any order.
- To be this flexible, there should be *no set-up time* in preparing for a new load. Tooling should either be available and resident in the process or be able to be delivered automatically from a master crib.
- *A program* to perform the task *should be downloaded* to the machine from area computers *as the iron is arriving* to be processed.
- *Quality should be measured as part of the process.* If a problem arises, the machinery and flow should stop. This is a major culture change from the current practice of continuing production and reworking parts at a later time.

G. Synchronizing Work Flow

When flow is synchronized, *all parts enter the process at the right time and are finished just prior to when they are needed at assembly.* Watching synchronous flow is like observing a ballet. There is continuity and harmony as material moves at an even pace, with everything happening at the right time and in the right order. Achieving this level of synchronization requires an invisible flow of information that parallels the visible flow of parts.

III. MATERIAL HANDLING AND SCHEDULING PRINCIPLES

A. Computer-Controlled Material Handling

Computer-controlled material handling is a critical aspect of a continuous flow factory. Material handling may be done with human-operated fork trucks, automated guided vehicles (AGVs), shuttles, or more sophisticated automotive-type overhead conveyors. The type of equipment is not important. *What matters is that the material handling equipment is controlled by area computers.*

In a continuous flow factory, material handling transactions happen quickly and frequently, and there is no inventory. Parts are processed in the order in which they arrive on the conveyor. They must be completed and moved to the assembly just ahead of when they will be needed. Timing is critical to synchronize parts processing with assembly needs. The only way to assure that parts present themselves for processing in the right order and at the right time is to control material handling by area computer.

When humans control the movement of material, flow is interrupted. People cannot keep pace with the number and rate of transactions that must take place. They are forced to squeeze in emergency orders. This disrupts the flow, causing problems and shortages and creating the need for large inventories which add cost and inefficiency.

Computer-controlled material handling is essential to synchronize flow in a continuous flow factory. The world's most successful manufacturing operations have it, even in very small plants. Many managers who do not understand the importance of integrated material handling delay its implementation or eliminate it altogether in an effort to conserve capital. As a result, the full benefits of modernization are not realized. As discussed in Chapter 2, in a typical modernization project, integrated material handling represents 20% of the total investment and produces 40% of the benefits. Therefore, if capital is scarce, it is preferable to buy fewer machine tools, but implement them fully with automated material handling and area computers.

B. Intelligent Sequencers

Sequencers are storage devices. They play a critical role in the continuous flow factory. Sequencers have different physical forms, ranging from a simple rack with lifting devices to an Automated Storage and Retrieval Systems (ASRS). The level of automation is not important. What is critical is that the sequencers have intelligence.

Sequencer intelligence creates flow and synchronizes it between processes. The logic is created by software located in the area computer. This software commands the sequencers and regulates movement of material through the mini-factory. The functionalities will be reviewed with systems in Chapter 7.

1. Sequencers regulate flow

A sequencer regulates flow in the shop like a dam regulates flow in a river. Different portions of a river flow at different speeds. Dams compensate for the differences. In the factory, parts are processed at different rates and in different quantities. For example, in the press shop parts move quickly and are processed in large batches. But in the weld area, they are processed at a much slower rate. Obviously, the press shop can produce parts faster than they can be consumed in the weld area, so a sequencer is needed to regulate flow between the two operations.

2. Sequencers schedule work

The second function of an intelligent sequencer is to perform the entire scheduling task for internal and external suppliers. This creates flow. The sequencer monitors parts consumption and signals (or triggers) internal and external suppliers to replenish parts as needed, in the proper quantities, just-in-time. The sequencer pulls the trigger early enough to give the internal or external supplier sufficient lead time to process and deliver the parts.

3. Use assembly-driven sequencer logic

Sequencer logic is assembly- or customer-driven. Assembly requires parts which must be retrieved or consumed from the sequencer. This is the first element of the logic. When assembly needs are satisfied, the second element of the logic is activated. A pull-trigger signals internal and external suppliers to produce and deliver just-in-time a new load of parts that will replenish the sequencer. The third element of the logic enables suppliers who receive the pull signal to have enough lead time to process and deliver the load just before it is needed in assembly. This requires that each part be produced in a load size that will meet assembly's needs during the time when a new load is being processed.

 This three-fold logic is similar to that used in supermarkets. The difference is that parts need to be processed before they can be delivered. This is the main challenge of sequencer logic in a continuous flow factory.

4. Use "first in, first out" processing

When rough or finished parts arrive from suppliers, they are delivered in loads to the sequencer. As production begins, a rough load is pulled from the sequencer and delivered to the appropriate operation (e.g., machining) where the parts are processed in the order in which they are received. We call this "first in, first out" processing. It requires flexible automation which will be described in Chapter 6. The flexibility makes it possible for the machinery to process parts in any order. Many different parts must be processed during a day. With flexible automation, the sequence in which they arrive at the process has very little effect on the timing of their completion.

 First in, first out processing can be compared to the take-off process on an airport runway. All jets get in line and take off in order. There is very little effect on the overall timeliness of a jet, as long as it takes off in a window of time. Likewise in the continuous flow factory, as long as loads take off from the sequencer in the right window of time, processes are completed in a timely manner. There is very little congestion or scheduling conflict because the action is taking place at the part number level in a manageable-size load.

5. Pull trigger simplifies scheduling

The pull trigger logic is the fundamental cause of enormous simplification in the scheduling of a continuous flow factory. The sequencer performs the entire scheduling task for the shop and external suppliers. Two conditions make it possible for suppliers to execute work in the order in which they receive pull trigger signals.

- The pull trigger signal is *random* because parts are replenished as they are consumed. When a load of parts is retrieved for use in assembly, the sequencer commands the internal or external supplier to produce a new part load. Scheduling production one load at a time avoids surges and inefficiencies.
- The processes are *flexible* and can complete part loads in any sequence in which they arrive. This permits flow to move continuously with no bottlenecks.

The importance of the sequencer cannot be overstated. *It schedules and manages the entire production of internal and external suppliers—triggering a continuous chain of signals which creates a continuous flow of iron.* The sequencer is the masterpiece of logistics technology and the "brain" of a continuous flow factory.

6. Role of the intelligent store algorithm

The logic that allows the sequencer to perform its vital role is called an "intelligent store algorithm." It is shown in Figure 4. This algorithm is similar to the minimum inventory algo-

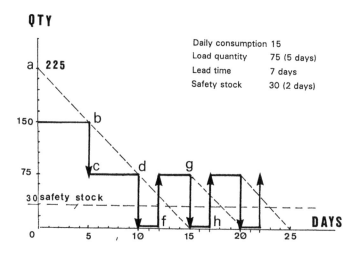

Figure 4 Sequencer algorithm.

rithm. To understand the algorithm, it is helpful to review a number of definitions.

- *Load* is the quantity delivered from the internal or external supplier. Load size is a function of daily consumption. A five-day load is used in the example.
- *Lead time* is the time it takes internal or external suppliers to process and deliver a load. The example uses seven-day lead time.
- *Safety stock* is the minimum quantity assembly carries at the time the sequencer receives a new load from the internal or external supplier. It is there in case assembly consumes more parts than planned. The example uses a two-day safety stock.
- *Average consumption* is the daily quantity planned for the assembly line. In this example, average consumption is fifteen.
- *Trigger* is the signal sent by the sequencer to the internal or external supplier, requesting a replenishment of one load.

The sequencer algorithm functions as follows:

Point	Day(s)	Status
a	0	225 units on hand (75 at assembly, 150 in the sequencer).
ab	0–5	Load of 75 units at the assembly line is reduced to 0.
b	5	Load #0 (75 parts) is sent to assembly. Sequencer inventory reduced to 75. Assembly inventory returns to 75. Sequencer sends pull signal #1 to replenish load. (Load will be delivered in 7 days).
bd	5–10	Assembly inventory reduces to 0.
d	10	Sequencer sends load #1 to assembly. Assembly inventory returns to 75. Sequencer inventory reduced to 0. Sequencer sends pull signal #2 to replenish load.
f	12	Load from pull signal #1 arrives. Sequencer inventory returns to 75.
g	15	Sequencer sends load #3 to assembly. Assembly inventory returns to 75. Sequencer inventory reduced to 0. Sequencer sends pull signal #3 to replenish load.
h	17	Load from pull signal #2 arrives. Sequencer inventory returns to 75.

This process is continuous, with the sequencer dispatching loads to assembly just-in-time and ordering replenishment loads from internal and external suppliers, allowing them to execute the order within their regular lead time.

Sequencer technology is known. What is new is allowing the sequencer command the entire scheduling of the plant. The sequencer and its logic are the "caddie master" of the factory. Using tee times, they synchronize the game in such a way that all golfers complete their rounds on time.

C. MRP II

MRP II (Manufacturing Resources Planning II) is a highly respected software package that is widely used for material procurement and factory scheduling. There is role for MRP II in the continuous flow factory—not as a daily scheduling system, but as a long-term planning system and supervisor of the pull trigger.

1. MRP II limitations for daily operations and scheduling

Daily operations in a continuous flow factory are driven by a fast, short-term execution system located in the area computer. This system provides the logic for sequencers and their interface with material handling systems and suppliers.

The fundamental reason why MRP II would not be effective as a daily scheduling system is that is *pushes* iron rather than pulling it. MRP II issues orders for parts lists (or bills of material) for a week or month of need. These are typically large-volume orders. As the orders are executed, conflicts arise, unplanned delays occur, and work flow becomes congested. Scheduling changes constantly as suppliers respond to a continuing state of emergency. To resolve scheduling conflicts, suppliers often start jobs earlier, increase lead time, and build in-process inventory—which is exactly what a continuous flow factory is striving to eliminate.

A second reason why continuous flow factories do not use MRP II for daily scheduling is its imprecision. If MRP II were used for scheduling, it would issue work orders based on stock record inventory computer files. These files are adequate for valuing inventory, but they are not accurate enough for factory scheduling. The errors they contain (due to data entry, counting, or processing problems) will have a negative effect on order quantities and completion dates, and ultimately disrupt flow.

A continuous flow factory needs an execution system with more speed and precision than MRP II can provide. The pull-trigger concept, which is the basis for the sequencer intelligence described above, serves as an "electronic Kanban." Its speed and accuracy make it the scheduling system of choice in many continuous flow factories.

2. MRP II strengths as a planning tool

The short-term horizon of the pull-trigger execution system is a disadvantage when it comes to planning. Although it is valuable for scheduling, the pull-trigger does not help plan rough material lines of supply, capacity, staffing, or other factors. This is where MRP II is needed.

MRP II is based on an overall program of production quantity (Q) and a list of parts needed for each product (P). P × Q = Requirement. This requirement is converted into purchase orders with broad delivery dates (by month) for a list of parts. This allows the company to secure lines of supply for rough and purchased finished needs. Requirement is also converted into manpower needed to achieve production.

MRP II plays an effective role in higher level planning. Located in the host computer, it issues purchase orders and establishes prices, permitting the company to manage capacity, employment, and lines of supply.

3. Using the pull-trigger and MRP II

The two systems work together in a continuous flow factory, with MRP II serving as the planner and the pull-trigger and sequencer serving as the expediter. MRP II identifies parts consumption above the original plan (griefs). When griefs arise, the pull-trigger execution system prevails—at least in the short term. If, for example, the execution system asks for

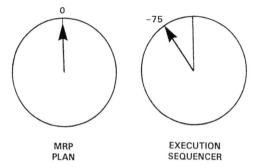

Figure 5   MRP grief.

a 75-part load, and the MRP reports that no parts are needed (Figure 5), demand should be satisfied first—then the two systems should be reconciled. Experience indicates that when the two systems conflict, the execution system is correct about 80% of the time. Griefs are frequently caused by entry errors or technical problems. For example, parts may have been scrapped, but the entry has not yet been made in the MRP II stock inventory record. Or an assembler is using two gaskets, one on top of the other, while the engineering bill of materials calls for only one. In this case, if the assembler is right, the bill of materials and design should be changed. Otherwise, the assembler should be instructed to follow the bill of materials to prevent future griefs.

To summarize, the execution system, located in the shop area computer, drives the sequencer and its interface with material handling systems and supplier just-in-time delivery systems to create the flow. MRP II software, located in the host computer specifies only broad delivery periods, controls order quantity and price, and helps reconcile actual consumption with the plan as it supervises the execution system. Both systems work in harmony and must be reconciled in a timely manner.

4. Real world problems

The combined use of MRP II and sequencer intelligence can adapt to problems in the real world.

- *Low quantity.* The algorithm in Figure 3 can use any quantity. If the average consumption is one per day, a load of seven days' usage would be seven parts. If the average consumption is three per day, a load of seven days' usage would be 21 parts. Lead time can be one day or 10 days.
- *Schedule reduction.* Both MRP II and the sequencer responses are amended if the schedule changes. If the schedule is cut, the assembly line must follow the new pace immediately. As parts are retrieved at a slower pace, the sequencer algorithm slows down replenishment. Rescheduling of internal and external suppliers happens automatically. The grief with the MRP II (Figure 5) will show very soon as MRP II will want more parts than needed. The MRP II will be amended.
- *Schedule increase.* If the schedule is increased, the assembly line should not be programmed with the increase until the sequencer has received new loads. A new amended MRP II is needed reflecting the lead time for the schedule increase. A simulation is possible with the algorithm in Figure 4. If we retrieve the new daily quantity, it creates a list of griefs which can permit us to expedite material for the schedule increase. It provides dock dates and load requirements.
- *Supplier delays.* This is a weakness of any just-in-time factory. Delivery delays cannot be tolerated. We should not create too much safety stock with the algorithm in Figure 4. If we establish a two-day safety stock, the supplier will soon understand the tolerance of the system and will use it. Suppliers should be cer-

tified for quality, quantity, and timeliness to assure a high-performance just-in-time factory.

For a more detailed discussion of material handling and scheduling principles, refer to U.S. Patent 5,193,065, which is assigned to Caterpillar Inc.

IV. QUALITY PRINCIPLES

A. Eliminating Quality Control Bureaucracy

Many traditional factories employ a large staff to serve as a quality police force. These people assess production quality, perform quality checks, sort out good and bad parts, and write scrap or rework tickets. They also monitor quality indicators, calculate quality ratios, and present many reports to management. This approach looks effective on the surface, but it has many limitations.

- Its focus is enforcement, rather than building quality into the process.
- No one is fully accountable for quality. Conflicts arise between the manufacturing organization and the quality police who share responsibility and accountability for quality.
- Operators may lack the tools to improve quality, yet they are punished when their production does not meet quality standards.
- Operators, who tend to view the quality police as a threat to job security, may choose to hide defects, rather than identify and resolve problems.
- Ratios, statistics, and reports rarely identify the cause of a problem. They merely quantify the number of quality problems an organization is having.
- Higher level management is not in a good position to solve conflicts at the part-number level. This is precisely where quality problems arise.

For these and other reasons, most modern continuous flow factories are dismantling their quality control bureaucracies and transferring accountability for quality to the operators.

B. Accountability for Quality

Quality control in a modern continuous flow factory takes place at the operator level. It is based on the premise that people take great pride in their work and have a strong desire to do a high-quality job. Companies capitalize on employee pride and commitment by making all people fully accountable for the quality of their own work. This approach has been prevalent in Japan for many years, but is now being used successfully in other areas of the world.

In order for this approach to be effective, management must create an environment conducive to quality. In such an environment:

- There is no quality police force.
- Quality is truly management's number one priority.
- Processes are capable.
- People have the tools and training that permit them to check production as it progresses, not after the fact.
- Flexible automation is used, freeing up operators to supervise their machinery and monitor their processes.
- A small number of technical experts is available to help operators with sensitive processes, machine programs, maintenance, tooling, and other potential problems.
- Quality certification is widely used. Candidates for certification include processes, individual machine operators of assemblers, work groups producing subassemblies, and entire factories.
- Employees are rewarded for quality achievements. Cash rewards can be difficult to administer, but other effective options include group recognition events, stories in employee publications, bulletin board displays, diplomas, plaques, visits to customer sites, or allow-

ing employees to serve as tour guides for customer visits.

C. Stopping Flow

In a continuous flow factory, there is no conflict between quantity and quality. Quality always prevails. There should be no flow unless quality parts are being produced. This philosophy represents a major departure from the common practice of producing poor quality and fixing it later. Management must make quality the number one priority and give employees the authority to stop production when quality problems arise. If production is stopped, management must act quickly, bringing together all available resources to solve the problem.

Despite the commitment to stop production for serious quality problems, there may be a few rare situations in which work must continue after a problem has been identified. In these cases, production with quality problems must be segregated from the main flow. If rework areas are to be used, the layout must be amended and provisions must be made to allow flexible machining systems to unload parts and route them to the correct location. There is no room in the layout to store a large quantity of rework. Therefore, if the fix is not fast enough, flow will stop.

D. "Poka Yoke"

During our visit to Citroen in Metz, France, we were surprised to see the Japanese phrase "Poka Yoke" displayed prominently throughout the plant. We were told the translation for Poka Yoke is "foolproof." It is a philosophy inspired by Toyota and includes a host of programs aimed at achieving zero defects.

Poka Yoke designs are foolproof because they have characteristics (like asymmetrical hole covers) that make it impossible to be assembled wrong. Poka Yoke assembly procedures are foolproof because each operator is responsible for checking the work of the previous operator, plus doing his own job. Poka Yoke product reliability is foolproof because all designs and processes are focused on eliminating assembly errors. The concept is simple, but the results are powerful. Designing products with the objective of easy assembly is a foolproof way to reduce costs and improve quality.

E. Process Capability and Process Control

It takes quality parts to achieve quality at assembly. Producing quality parts is a function of process capability and process control. To assure process capability, the product design and manufacturing process must be in harmony. The design should demand realistic tolerances that are achievable by a state-of-the-art process. There is a tendency in engineering to set tolerances too low, believing that manufacturing will not deliver what has been asked of them. The strategy is to ask for more than they need so they can get what they need. This is wasted effort. Manufacturing should comply with the blueprint—no less, no more—and make available to engineering the full capabilities of the process.

The target in Figure 6 illustrates the concept of process capability. The bullets are dispersed within a circle called "capability." This dispersion represents the full range of process tolerances available to engineering. In this case, we

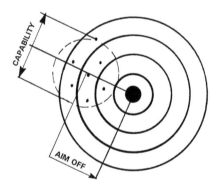

Figure 6   Process capability—aim off.

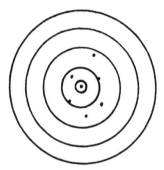

Figure 7 Process control—good aim.

cannot hit the target because the aim is left and high. To reach the target, we must change the aim. If the dispersion is too wide, we must get another gun (select a new process). Figure 7 shows the same capability, but the aim has been corrected through process control. As the diagram illustrates, if the process is capable and the aim is good, quality is on target. Process capability is achieved by simultaneous product and process engineering. Process control is achieved through quality management in the shop. A continuous flow factory must have both.

Many methods are used for controlling processes. The most simple approach is to use a manual gauge to check critical dimensions. By plotting the results on a chart, the operator knows what is happening, sees the trend, and can take corrective action when necessary. Another approach is automatic gauging using the numerical control of the machine to probe the dimensions and display the results. Probing requires machine movement and slows down production. Therefore it is only used for critical applications when it is the best alternative.

Process capability and process control can be measured in a single formula called CPK which is very popular in the United States. This formula involves the "bell curve" statistical law. When CPK is 1.33 or higher, the process is capable and under control. CPK is explained in depth in *Statistical Quality Control* by E. L. Grant and R. S. Leavenworth (McGraw-Hill Book Co., New York, 1988).

The supreme weapon for process control is a coordinate measuring machine (CMM). This is a programmable three- or four-axle, high-precision measuring device. It helps assure that machines have been set right and remain on their correct foundations. A CMM is expensive. It is also slow and checks quality after—rather than during—production, but it is the best technology available for process control. The machine should be located in a position that allows random access to it by shuttle.

V. SUPPLIER PRINCIPLES

A. Supplier Certification

Supplier certification is a critical aspect of continuous flow manufacturing. Certification involves setting stringent quality standards and working with the internal or external supplier to meet those standards. Certification is an excellent way to assure that suppliers' organizations are focused on quality and suppliers' processes are capable and under control.

B. Packaging and Bar Code Technology

To automate receiving, traditional activities such as parts counting and identification must be eliminated. This can be done with packaging and bar code technology. Packaging is a technique in which suppliers deliver a fixed number of parts in a box or on a pallet. Each of these "packages" is a fraction of a load, meaning a load may consist of ten to twenty packages. Each package has a bar coded identification label.

Trucks arrive at a dock and are unloaded by fork lift operators. Packages are placed on a receiving conveyor. There the bar code labels are screened, and the parts enter the system without any other intervention. As the truck is unloaded, the receiving operator checks a screen to determine how many packages constitute a load. Since the factory is controlled by loads, no partial loads should enter the system.

Packaging requires teamwork between the manufacturer and supplier. The supplier is responsible for packaging in the right quantity and in standardized boxes and pallets. The manufacturer should make the quantity and size requirements as cost effective for the supplier as possible.

Packaging is essential in a continuous flow factory. At least 90% of all purchased finished parts must be packaged in order to maintain efficient and cost effective flow.

C. Consolidating Traffic

When a company has several plants, each receiving small loads, it could be more economical to have parts delivered to a strategically located warehouse where they are consolidated by plant for delivery in full loads. This can shorten delivery distance for suppliers and reduce traffic at the plant. Two important ground rules for these "pass-through" warehouses:

- No repackaging is allowed.
- Waiting time for consolidation is limited to a maximum of two days.

D. Point of Use Delivery

In some situations, truck drivers will become familiar with the layout of the plant and will be able to deliver loads the point of use in the factory. This lowers material handling costs and reduces lead time. The truck drives the extra distance at no cost and unloads at docks situated near the sequencers (see Figure 3).

E. Other Logistics Concepts

There are other innovative ways to streamline logistics. One is to use a third party distribution company. With this approach, the manufacturer buys from the supplier on an annual basis—negotiating the price, setting quality standards, and so on. But it is a nearby distribution company that actually monitors consumption in the plant, orders parts delivery from suppliers, and handles all storage and disbursement functions. This approach is widely used in the automotive industry as a means of lowering costs, improving efficiency, and reducing lead time, especially for low-cost parts like hardware. It has shown good results for lower volume businesses like construction equipment manufacturing.

Even more advanced logistics concepts are being used by some auto makers. For example, at General Motors in Buick City, suppliers of certain parts receive orders annually, monitor production, and feed the line as needed. Some parts are actually delivered to the assembly line every few hours. The supplier owns the inventory until the vehicle is produced and is paid by the number of vehicles completed. This approach is the same as the method used for selling candy bars and soft drinks in the plant's vending machines. Similar techniques are being used effectively in plants in the United States and Europe. Companies that use this approach can reduce the number of part numbers in their systems by one-third or more, reducing clerical effort and administrative costs.

VI. HUMAN RESOURCE PRINCIPLES

A. Employee Pride

There can be no quality in a plant unless the workforce is highly motivated. One of the fundamental sources of motivation is pride. It is difficult to generate pride in a traditional factory where:

- The pipeline assembly line is being pushed beyond its limits.
- Too many people have been employed in an effort to achieve production.

- Jobs are narrowly defined and virtually all authority rests with management.
- Work flow is congested and pressure is high.
- Meeting production rates is a higher priority than producing quality.

In a continuous flow factory, pride is generated in a number of ways.

- Employees have the right tools for the job.
- Job content is broader, and work requires less physical effort, but more thinking and reasoning.
- Operators have ownership in a total process and a finished product.
- Quality training is a priority, and operator certification is a common practice.
- Employees are involved in decisions that affect their work lives. They participate in the design of their work stations.
- Employees understand the roles they play in the overall process and have exposure to their internal and external customers to display their quality efforts.
- Employees have more autonomy, additional responsibilities and the authority to stop production for serious quality problems.
- Managers create an environment in which quality is truly the number one objective.

Factors that threaten morale in a continuous flow factory include:

- Insufficient training.
- Inadequate maintenance of machines.
- Job classifications that do not accurately reflect the complexity of the job.
- Too much emphasis on unattended machines, which results in the perception that people are unnecessary in the factory.

When these factors are addressed and resolved by management, the continuous flow factory is a positive, productive, and high-quality work environment.

B. Communication Between Operators and Programmers

To get maximum uptime—and therefore top production—from flexible machining systems (FMS), there must be ongoing communication between operators and programmers. We often make the mistake of assuming that because senior operators do not understand how computers work, they cannot contribute to FMS programming. The fact is, their years of experience cutting, turning, milling, and drilling iron manually make them a valuable resource to programmers. Operators can help programmers better understand the iron cutting mechanism. A good operator can convince a programmer to focus on uptime rather than cycle time by demonstrating that a 3% longer cycle with 95% uptime yields more production than a shorter cycle with 60% uptime.

C. Reducing Turnover

Job stability is vital in a continuous flow factory. A company cannot afford to lose an operator after investing in months of training, sending the person to the machine builder, and enduring a painful ramp up period. At Caterpillar, we had a problem with an FMS and as we attempted to resolve it, we discovered that we had changed operators seven times in a six-month period. Obviously, job turnover was contributing to the situation.

Management must take steps to assure job stability. Two suggestions are either to create protected job classes or to put two operators on an FMS and allow only one position to move. Whatever the solution, turnover must be reduced. In Japan, it is not uncommon for employees to spend ten years on the same job. Even in Europe, there is more stability. Employee turnover is a major source of inefficiency and must be addressed by management.

D. Budgeting for Training

One of the most frequent complaints of operators is insufficient training. Training should be a priority and be budgeted as part of start-up costs. Although basic training in English and math skills is important, the highest priority should be given to specific, focused training on the type of machinery the employee will have to use. If an FMS is being purchased, it may be useful to buy one machine center ahead of time for in-shop training of operators, maintenance people, programmers, tool grinders, and others. For more complex machines, it may be necessary to purchase a simulator for training.

VII. SUMMARY

Key points of this chapter:

- The principles of continuous flow manufacturing were acquired by visiting more than 100 manufacturers in the United States, Japan, and Europe.
- A modernization plan based on automation alone will not produce the results that can be achieved by a plan based on all the principles presented in this chapter.
- The principles of continuous flow manufacturing can be divided into five major categories: basic, material handling and scheduling, quality, supplier, and human resources.

VIII. FUNDAMENTAL PRINCIPLES

F5. Material flows in one direction.

F6. Loads of parts are started and completely finished within one machine process and are machined in the sequence they arrive in the process loading area.

F7. Parts flow by using computer controlled material handling. Management's job is to maintain flow—not cause it.

F8. Flow is synchronous. A load of parts must be started at the right time to be completed just ahead of need in assembly.

F9. Machines are flexible and do not need set up time to start and complete the loads assigned to them.

F10. Machine programs and tooling are resident or can migrate quickly.

F11. Quality management is integrated into the process.

F12. Material handling devices maintain the flow. They can be manual, but they are all monitored by computer, which causes and synchronizes the flow.

F13. Sequencers are needed to regulate flow, compensating for differences in the speed or volume of adjacent processes.

F14. The sequencer system and logic create the flow as they ask to be replenished as needed. The sequencer pulls a trigger for more iron, giving internal or external suppliers sufficient lead time to respond.

F15. Sequencer logic within area computer software controls the execution system. The execution system orders parts within a precise window of time to allow just-in-time delivery by internal or external suppliers.

F16. Manufacturing Resources Planning (MRP II) is reserved for long-term shop and supplier planning of rough material, human resources, and so on. It should not be used to schedule the shop, but to supervise the execution system.

F17. There is very little or no buffer inventory.

F18. If there is a quality problem in machining or assembly, flow stops.

F19. Machines, equipment, and all processes have the capability to meet design tolerances. Simultaneous engineering has insured that design and process are compatible.

F20. Machinery has process control for production to be within tolerances. Process control can be assisted by manual gauging, automated gauging, or coordinated measuring machines.

F21. Certify internal and external suppliers for good quality.

F22. Suppliers deliver parts in standardized packages which contain known quantities and have bar code identification labels.

F23. Supplier delivery is performed at the point of use in different zones of the plant.

F24. External logistics may be provided by third party organizations.

F25. Suppliers may have access to assembly locations and replenish bins as needed, being paid by loads without paperwork.

F26. Employee pride is generated through greater autonomy, broader-scope jobs, and personal accountability for quality achievement.

F27. Job content is broader, permitting more efficiency, competence, and coordination. It requires extensive training.

4

Automation for the Continuous Flow Factory

I. INTRODUCTION TO AUTOMATION

The philosophy of automation in a continuous flow factory was introduced in Chapter 2. To recap, automation should be:

- *Reasonable*, using proven technology at the peak of supplier competence
- *Integrated* with logistics and material handling systems
- *Flexible* enough to completely finish parts in the order in which they present themselves
- *Deployed* within an efficient flow layout

There are four basic types of automation in the continuous flow factory arsenal:

- *Manufacturing cells* are designed for low-volume production and are flexible enough to process families of parts.
- *Transfer lines* are relatively inflexible. They are best suited for high-volume production of a single or very small number of parts.
- *Cellular systems and flexible machining systems (FMSs)* are computer controlled machining centers that process groups of parts with similar designs.
- *Flexible assembly and welding systems* are computer controlled processes that permit a variety of products to be assembled (or welded) and in any order.

This chapter discusses the four types of automation and provides examples of each. Some examples contain documentation reproduced with permission from machine equipment and system vendors. The examples are included to illustrate concepts. They should not be interpreted as advertisements for the products mentioned.

II. MANUFACTURING CELLS

Three types of manufacturing cells will be reviewed in this section:

- *Simple cells*, which are the generic application of cell principles and are very popular around the world
- *Automated storage cells*, because they utilize automated storage and retrieval system (ASRS) technology to accelerate flow
- *Forced flow cells*, as they are popular in Japan and emphasize machine utilization

A. Simple Cell

A simple cell is a group of all the machine tools and related operations necessary to completely process a family of parts.

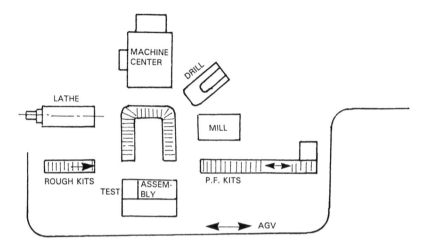

Figure 8 Gear pump cell.

A cell may include subassembly operations. Figure 8 shows a simple cell that machines, assembles, and tests pumps. The cell includes:

- A machine center to machine the pump body and cover
- A tracer lathe to turn the shaft
- A mill for machining the key on the shaft
- A drill for making lube-oil cross holes

The machine center can store eight different programs for four pump bodies and four covers. The mill and drill have no setup time, and all tooling stays in the cell. The cell is not computer controlled, but there is a terminal in the cell which links it to a rough material sequencer, an assembly sequencer, and computer controlled material handling.

Work flow starts when the assembly sequencer demands a load of pumps. Materials are pulled from the rough sequencer and delivered to the input conveyor on the left by automated guided vehicle (AGV). The material arrives in forging and casting kits. Purchased finished items such as gaskets and hardware are also delivered by AGV in kits to the conveyor on the right. The operator loads the machinery manually, carrying a part from the conveyor in the center to a machine, then returning it to the conveyor where it awaits the next operation. All parts are completed and deburred, then assembled into pumps at the assembly bench. Testing takes place at the test bench. This approach is effective because parts are completely processed, assembled, and tested with very little inventory. There is no paper involved —just a pull trigger from the assembly sequencer that signals the delivery of materials to the cell on time and in the right quantity. As long as materials arrive as scheduled, the operator can handle all aspects of pump production. He is fully accountable for the entire process and requires very little supervision. As a result, job satisfaction and motivation are high.

There are two drawbacks of a cell like this. First, it requires a lot of floor space. And second, utilization is low because the machines must be activated by the operator who is busy doing many jobs. For these reasons, it is best to use cells when extra floor space is available and when older, manually controlled, and stand-alone machines are being retained.

B. Automated Storage Cell

An automated storage cell (Figure 9) is similar to a simple cell, but it uses an automated storage and retrieval system (ASRS). All machine tools (A through L) are laid out around two L-shaped conveyors. These conveyors receive rough parts in tote boxes and interface with the ASRS. All machines and the ASRS are monitored by a cell computer. A load of parts arrives and is routed to the first operation (at Machine A, for example). When the operation is complete, the load, made up of one or more tote boxes, re-enters the ASRS. The computer knows that the first operation has been performed and will schedule the second operation (at Machine D, for example) as soon as possible. This forces completion of the part in mini-

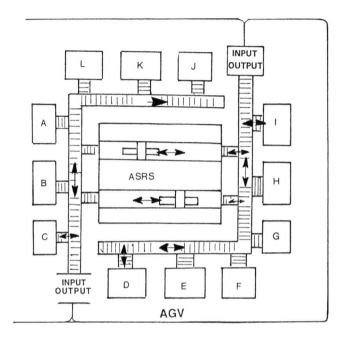

Figure 9 Automated storage cell.

mum lead time. All subsequent operations are scheduled quickly to keep the iron moving.

The Sikorsky Helicopter plant in the United States has used automated storage cells successfully. This approach accelerates work flow, reduces lead time, and represents a good first step toward a continuous flow factory. However, it is still a "push" type factory where partially processed iron waits to be scheduled between operations.

C. Forced Flow Cell

Forced flow cells are popular in Japan. They use no computers and are labor intensive. We saw excellent examples at Nippon Denso and Kayaba.

Figure 10 shows a forced flow cell. It resembles a tunnel, with machines laid out alongside one another on both

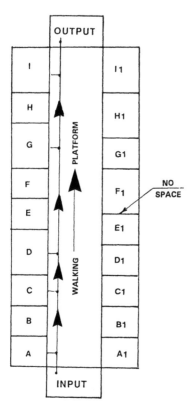

Figure 10 Forced flow cell.

sides of a walking platform. Machine set-up is permanent. Flow moves in one direction—from the bottom of Figure 10 to the top. Rough materials are received at the input table in order of the need for finished parts. Finished parts exit from the output table. Most parts require four to five operations.

Parts are processed on a first in, first out basis. An operator's job involves carrying the part to the appropriate machine, putting it on the table so it will be ready to load when the previous cycle stops, unloading the machine at the end of the cycle, loading the next part which waits on the table, and carrying the finished operation to the next machine. This cell processes parts like an FMS, but material movement is all manual. Productivity can be extremely high in a forced flow cell. Total cycle time is the sum of walking time (time spent moving between machines) and spindle time (the time in which the machining operation is actually being performed). Since spindle time is fixed, the only way to reduce cycle time is to cut walking time. We saw operators in Japan running on the platform. Some were able to achieve one-minute cycles, with total cycle time only 10% above spindle time. Although this display of productivity was impressive, it would be difficult to achieve similar results in most factories.

D. Cell Utilization

As mentioned previously, utilization is low in most cells. Machines often sit idle for long periods of time waiting to be activated by an operator. In a study of utilization, Giddings & Lewis, the largest machine tool and systems manufacturer in the United States, determined that utilization rates (the ratio of spindle time to total available time) are directly related to the level of automation. The results of the Giddings & Lewis study are shown in the following table:

Average Utilization (Spindle Time as a
Percent of Total Available Time)

Stand alone machine	12%
Computer controlled machine center	35%
Computer directed cell	80%
FMS	90%

Knowing the utilization rate that can be expected from automation is extremely important because it affects the return on investment. When utilization is high, fewer machines are needed to meet capacity requirements. Fewer machines will result in a better return on investment.

The example shown below compares the economics of purchasing an FMS versus several machines that will be grouped in a cell. In both cases, 5.4 units of capacity are required. The FMS has 90% utilization, the cell 45%. As the chart indicates, to obtain an equivalent amount of capacity, the company can invest $5.7 million in an FMS or $7.4 million in four cells. Obviously, the FMS will have a better return on investment.

FMS Versus Cell
(Dollars in Millions)

Configuration 1: FMS

Six machines and pallets	$ 4.8
Computer	0.3
Shuttle	0.6
Total	$ 5.7

Utilization: 90%
Capacity: 5.4
(6 machines at .90 utilization)

Configuration 2: Cell

Four cells of three machine centers and tooling	$ 7.2
Two stand alone machines	0.2
Total	$7.4

Utilization: 45%
Capacity: 5.4
(12 machines at .45 utilization)

These numbers are overly simplified because there are other costs associated with an FMS. Pallets and tool magazines, for example, will cost more with an FMS than with a cell. But when all the costs are considered, if new technology is being acquired, an FMS will yield a better return than a cellular system. An FMS will also require less floor space.

III. TRANSFER LINES

A. Dedicated Transfer Lines

A dedicated transfer line processes a single part number, accepting very little variation of the part design, such as additional drilled holes. It is made up of machining heads for milling, drilling, tapping, or boring. The heads are laid out in different stations and a station can have one or several heads. Stations are connected by a conveyor. Parts move from one station to the next automatically and are fully processed when they reach the end of the line.

This technology is "dedicated" to a single part because each head is laid out in a fixed position relative to the design of the part. For example, all the drilled holes have a fixed dimension between axles. The spindles rotate (speed) and penetrate (feed) the iron with the single purpose of cutting. Motors are electrical or hydraulic. There is no need for a computer to control movement between spindles because their relative position is frozen, based on the part design.

A dedicated transfer line is an economical and productive choice when the part design will be stable for many years. But if the design changes for any reason, the transfer line is obsolete. Dedicated transfer lines have been widely used in the automotive industry where some parts have had very long lives. But in most industries today, including the auto industry, product designs are changing faster and more frequently, so there are fewer applications for this type of automation. The trend favors convertible or flexible transfer lines.

B. Convertible Transfer Lines

Convertible transfer lines can be changed as part design changes. Cross, a Giddings & Lewis company, uses modular design to allow faster, more effective conversion of transfer lines. Modules such as the head, cooling system, shuttle section, and electrical components, can be changed to accommodate new part designs. With modular design, Cross attempts to reuse 60% of the original asset and make the conversion in one to two months. These are ambitious—but admirable—goals, and if they can be achieved, the transfer line concept will be a more practical option in today's dynamic business environment.

C. Flexible Transfer Lines

A flexible transfer line uses standard machine centers with multi-spindle heads that can be changed by head changers in ten seconds or less. There is no set-up time between part changes. The use of different pallets and head changers makes this type of transfer line flexible enough to process larger families of parts. For example, a flexible transfer line can machine a batch of four-cylinder engine blocks, then six-cylinder blocks using the same bore diameter. It can also machine two different block designs.

Flexible transfer lines use more standard machinery like computer-controlled machine centers. They are closer to the FMS concept, but still dedicated to a small family of a similar pieces. Like all transfer lines, flexible transfers include one-of-a-kind machinery for unique parts. Flexible transfer lines run pieces in batches to reduce tool change time. A modern flexible line phases in a new batch of parts and phases out an old one without stopping for set-up.

There is complexity associated with flexible transfer lines—more traveling, tool changes, head changes, and program changes. And like the FMS, a flexible transfer line needs more controls and computers. Flexible transfer lines have been criticized for their single-spindle heads, slow speed, and lack of spindle rigidity. While these criticisms may have been true in the past, they are no longer valid. Today's heads can have multi-spindle drills or taps driven by the center spindle of the machine, allowing the flexibility of multi-spindle cuts. Spindles on current machines are also faster and more rigid than they were in the past.

Computer control has created a new dimension in transfer line versatility. Unfortunately, many companies, especially in the United States, are not taking advantage of this new dimension. Following is a side-by-side comparison of dedicated (or convertible) and flexible transfer lines.

Dedicated/convertible transfer lines	Flexible transfer lines
• High quantity	• Lower quantity
• One part number or minor variations	• One family of similar parts with more variations
• Proven technology	• Proven technology, but need excellent multi-spindle head changer
• Simple	• More complex
• Little software required	• Need computer network
• May be convertible for new parts (40 to 50% wasted)	• Convertible with new heads, only tooling wasted (20%)
• Rigid	• Rigid (hydrostatic) benefit of standard development
• One-of-a-kind machines	• Fewer one-of-a-kind machines
• Can use standard modules	• Uses standard machine centers
• Cannot be exploded, needs rebuilding	• Can be exploded, machine centers can be used in cells
• Popular in United States	• Well developed in Europe

IV. CELLULAR AND FLEXIBLE MACHINING SYSTEMS

Cellular systems are comprised of flexible machining centers assembled with other multipurpose machines. They process families of parts and are computer controlled for programming and tool management.

An FMS is a larger system with highly automated material handling and tool delivery. A network of computers controls the functions of an FMS, while operators load and unload parts and supervise the flow. An FMS has several like machining centers. Automated tool delivery makes it possible for the computer to direct each part to the first available machine. This random access to all available machining centers improves utilization and return on investment.

A. Cellular System for Fluid Pump Casing

Figure 11 shows a Giddings & Lewis cellular system (reference case number 3322). The system features a raw material ASRS which feeds a subassembly sequencer close to the final assembly area. This is a flexible system that machines 22 different parts. Cube size is limited to $4 \times 3 \times 2$ feet. The pumps machined on this system have been greatly simplified to improve manufacturability. They are made of split casings. Both halves of the casing are premachined on the system. Then a subassembly operation is performed, followed by finish machining back on the system.

The architecture of the system is simple and was designed by the vendor. A computer numerically controlled Numeripath 8000 and CM 9000 cell controller perform scheduling, program downloading, part routing, tool management, reporting, and AGV move control. The machinery consists of two MC 60 horizontal machining centers with 120 resident tools and load-unload rotary stations. Palletization by the operator includes a push-button ready message to the G & L CM 9000 which is interfaced with the AGV to transport the pallet to an open machine.

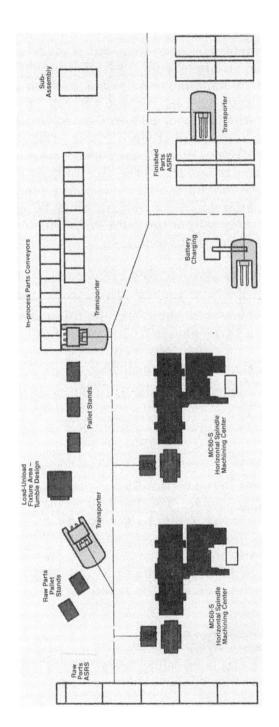

Figure 11 Cellular system—pump casing.

77

This is a simple cellular system, but it is far more productive than the primitive pump cell shown in Figure 8. Automated material handling and flexible machining centers enhance its productivity. And because the cell is computer controlled, utilization is about 80%, compared to 12–15% in the manual cell. Total productivity for this arrangement is approximately 15 times that of stand-alone machine tools. Inventory has been reduced by 50%. This cellular system is especially effective for the two-step machining process with subassembly in between. More volume could be accommodated by adding more cells of the same kind. Due to the two-step process, a full-scale FMS would not be appropriate.

B. Cellular System for Clutch Brake Components

Figure 12 shows another cellular system from Giddings & Lewis (reference case number 3303). In this example, the system evolved in three phases from a simple cell to a full-scale FMS.

Phase I is a simple cell and included:

- A vertical machining center with 80 tools and a tool changer
- A vertical turning center with 30 tools used as stand-alone
- A Giddings & Lewis computer numerical control

Parts are processed in the vertical turning center, then in the vertical machining center. The machine center possesses numerical control which positions the spindle and machine according to the program. Conventional forklift trucks are used for loading and unloading.

In Phase II, material handling was integrated, producing a cellular system. Additions to the cell were:

- One transporter
- Set up rotary stands
- One washer
- One G & L CM 9000 cell controller Digital DEC Micro VAX II

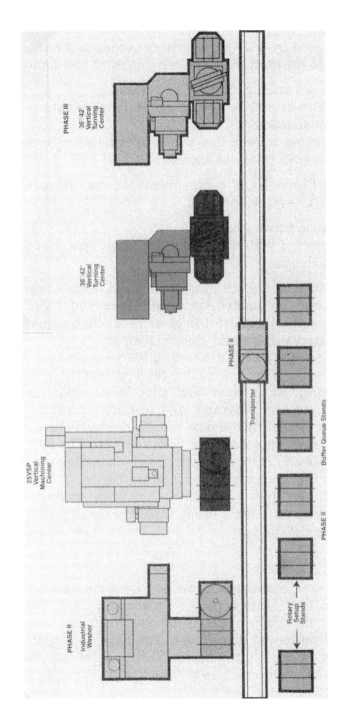

Figure 12 Cellular system—clutch brake.

Parts are loaded by AGV to the rotary stands and routed by transporter to the machines. The cell computer coordinates:

- Parts scheduling, routing, and tracking
- Downloading of the CNC program for each part
- Tool management
- Production information which appears on a terminal screen or in print-out format

During Phase III, the full-scale FMS was introduced. Additions to the system included:

- A second turning center
- Integration with area computer (with DEC Net Link and Ethernet)

When the full-scale FMS was introduced with computer monitoring and automated, integrated material handling, two operators could achieve the same production volume as twelve operators using stand-alone machinery.

Phasing in the introduction of an FMS can produce positive long-term results. It provides the opportunity to maintain production at a relatively high level while testing perishable tools and programs and training operators and maintenance people. By the time the entire FMS is launched, many of the problems will have been resolved. The key to a successful phase-in is to plan each phase—including the final one—before the implementation process begins. A comprehensive plan for layout, material handling, tooling, programs, and long-term labor needs should be developed at the very beginning of the process.

C. Full-Scale FMS

An FMS can be the supreme weapon for a continuous and synchronous flow factory. In the right application, nothing can surpass the FMS in terms of cost reduction, quality, output, and return on investment. The FMS is a relatively new concept in the United States. Prior to the late 1980s, demand for these systems was low. Large manufacturers preferred transfer line technology and medium-sized companies did not yet see the advantage of processing parts in families. Because the market was weak, machine tool vendors in the United States did not develop and promote good products.

In Europe and Japan, however, small, low-volume companies began to realize that they could achieve economies of scale by processing similar-type parts in larger groups. Some parts were grouped according to size; others by process or machining operation. This concept helped create demand for the FMS outside the United States. Now the practice of processing parts in groups has become more widespread in the United States. As a result, demand for FMSs has increased and machine tool vendors have greatly improved their product lines.

1. FMS selection criteria

FMS is not the best choice in every application. Many factors must be considered to determine the best technology for the job. Following are some of the key issues to evaluate when considering cellular systems and FMSs.

Cellular system	FMS
• Need to segregate iron by size (small, medium, large)	• Need to segregate iron by size (small, medium, large)
• Small family of parts (10 to 50 of one size)	• Large family of parts (100 or more of one size)
• Process continuity is key issue	• Utilization is key issue
• Process requires different types of machines, few machines of the same kind	• Process requires different types of machines, five or more machines of the same kind
• Tool management manages tool wear and changes, often need operator to serve machine	• Random machining and fully automated tool loading are justified by utilization gain and savings in tooling

Cellular system	FMS
• Cell control computer and simple systems do not require high-level specialist	• More sophisticated computer and complex system requires technical specialist, affordable only for large scale processes
• Can be integrated with area computer and ASRS	• Usually integrated with area computer, rough ASRS, assembly sequencer, and possibly to host computer
• Use of nearby coordinate measuring machine (CMM) is generally manual	• Can afford a captive CMM with automated access due to volume; another option is an automatic gauging cell with computerized tool compensation at the machine
• Tool setting is usually manual on demand of cell control, could have tool room scheduling	• Tool setting is manual with automated help, chip reader inserts in tool holder, continuous tool life log, and tool room scheduling
• Tool delivery is manual	• Tool delivery is mechanized by shuttles and manipulators to feed machine center magazine, less tooling is resident, large tool migration
• Simpler, easier to get started	• More complex, harder to keep simple and running, needs experience and expertise
• Can run 90% uptime within 6 months	• With phased introduction, can run 90% uptime at best after one year
• Need several cells to make a factory	• Entire factory can be one or two FMSs with many identical machines

- Potential for good productivity, cost reduction, utilization, and return on investment

- Potential for the highest level of productivity, cost reduction, utilization and return on investment

2. Managing FMS complexity

An FMS can be the best option in many applications. However, they can be very complex and must be carefully managed. Following are suggestions for managing the complexity of an FMS:

- Invest in top-quality programming. If you want production to flow like a river, you need excellent software and cannot afford to minimize the investment in programming.
- Get an FMS machine center early to develop perishable tooling and programming, train operators, and verify quality standards.
- Have the FMS operator and those responsible for FMS maintenance, particularly electricians and hydraulics technicians, participate at the supplier's assembly and test run.
- Use simultaneous engineering to harmonize the part design and its manufacturing process.
- Have the supplier design and execute the FMS foundations, including the chip evacuation system.
- Make suppliers responsible for pallet and durable tooling design, using your company standard to avoid creating new tools.
- Conduct a separate try-out for shuttle accuracy serving the pallet in and out of the machine, gantry robots, shuttle, etc.

3. Simple merry-go-round FMS

Although there is a degree of complexity associated with all FMSs, some systems are fairly simple. Many Japanese companies use a simple merry-go-round FMS that features standard machining centers and a loop conveyor with one loading and unloading station for palletization of parts. Programs are stored at the cell controller level and downloaded. There is no automatic tool management. All tools are resident and can be changed manually.

We have seen a simple Mitsubishi FMS that was started up in three weeks and ran short of iron because the programs had not been completed. Management had expected to have three months for programming. The Japanese service technicians went home and came back when the programs had been completed. Unfortunately, there are very few examples of this type of FMS in the United States. Many manufacturing engineers prefer complex technology, even when it is not necessary. The fact is, major gains in productivity can be realized by using the simple FMS to process high-volume, repetitive parts.

One cost effective way to meet high-volume production goals is to use three merry-go-round FMSs—one each for small, medium, and large parts. These highly productive and trouble-free systems can be worked in three shifts a day, and run unattended during lunch periods and for half a day on Saturday. This produces a very attractive return on investment.

4. Higher technology FMS

There are certain applications which are better suited to a higher technology FMS. One good example is the machining of hydraulic valves that will be used in a large family of products. There could be as many as 500 different part numbers—all similar, yet slightly different. This application could require ten or more separate cellular systems, which would be difficult to manage. A better solution would be to invest in a group of high-technology FMSs, supported by automated parts and tooling logistics systems. This is a highly advanced operation which could achieve 90% or higher utilization, reducing the number of machines needed and permitting an excellent return on investment.

An FMS alone cannot achieve this level of utilization. It must be supported by a full deployment of technology including:

- An automated parts logistics network
- An advanced internal scheduling system
- Integrated washing and measurement systems
- An automated tool logistics network

a. Automated parts logistics. Figure 13 shows a group of linked FMSs interfaced with an intelligent sequencer or ASRS. As discussed in Chapter 3, an assembly sequencer controls production by transmitting a pull signal to the FMS, telling it to begin processing a load of parts

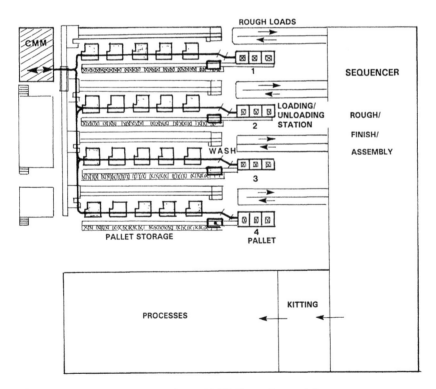

Figure 13 Higher technology FMS. Interface with sequencer.

needed by assembly. Iron for the load is pulled from a rough material sequencer and delivered by high-speed shuttle to one of the four load and unload stations at the FMS. There the operator manually loads the part into a pallet clamping device. Each pallet is capable of clamping several types of parts. After the material has been palletized, it is transported to the appropriate machining center and processing begins.

b. Internal scheduling system. It takes an excellent internal scheduling system to maintain fast, efficient work flow on a complex FMS such as the one shown in Figure 13. At any given time, each of the four load/unload stations may be receiving two or three different loads of iron, some partial loads, others full. The FMS may be machining up to 12 different parts at a time. And each part may need to be routed to more than one machine to be completed. Efficient pallet movement is essential to maintain top production in this kind of environment.

The FMS computer schedules and controls pallet movement. Each pallet is routed from storage (in the pallet magazine) to the load station; then through the machining and quality control processes; then back to the unload station where it will be cleaned, unloaded, and either loaded with another part or returned to storage.

It is important to have the right number of pallets to maximize machine utilization. For best utilization, there should always be a pallet loaded with parts waiting at the machine. If two machines are being used for one part, there should be at least eight pallets available: two at loading, two in transit, two waiting at the machine, and two working at the spindle.

c. Integrated washing and measurement systems. In the high-technology FMS in Figure 13, two important functions have been integrated into the total process: pallet washing and quality control. Pallets are washed at the load/unload station prior to being loaded with new parts.

Washing removes chips and permits more reliable loading and clamping of the part.

Quality measurement takes place at a coordinate measuring machine (CMM). Parts are transported to the CMM where critical dimensions are checked after the machining has been completed. If problems are detected, the CMM tells the FMS to stop production.

There is some risk associated with checking part dimensions after, rather than during, production. Machine tool vendors do provide probing devices to take critical measurements at the machining center. These probing devices can be useful with cellular systems or simple FMSs, but on a high-technology FMS such as the one shown in Figure 13, probing at the machine center is too costly because it takes machine time and slows down the production process. For a large, complex FMS, quality is actually managed through tool compensation and management. CMM serves as a quality insurance policy only.

d. Automated tool logistics network. In a group of linked FMSs, there may be 12 or more parts being machined at the same time. If each part uses an average of 10 tools, the system needs to have 120 tools at work and prepare 120 for the next job. To respond to this high demand, a complete tool logistics network is required. A tool logistics network is shown in Figure 14. It includes:

- Tool management to initiate tool change as wear takes place
- Automated tool delivery at the machine
- Tool scheduling and tool setting to prepare tools
- Tool compensation to manage quality through process control

Tool management. Most machine vendors determine the maximum number of parts that can be cut by one tool. The number is placed in memory, and a tool change is generated when the number is reached. This is called tool management. Tool management helps maintain process control

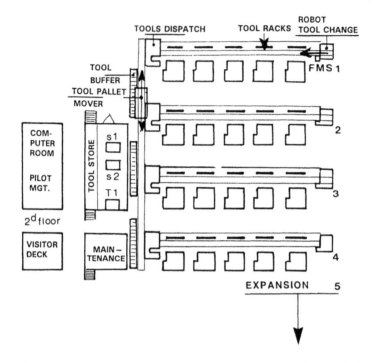

Figure 14   Higher technology FMS. Automated tool delivery.

and is critical to stabilizing production at a high-quality level.

Automated tool delivery.   For a high-technology FMS to achieve maximum utilization, there must be a system for automatically delivering the right tooling to the right machining center at the right time. This is called automated tool delivery. It requires a separate logistics system shown in Figure 14. The system includes:

- A tool store that includes automated tool setting
- A tool pallet mover to move tooling from the tool room to the tool pallet buffer and from the buffer to the tool dispatch carrier
- A tool dispatch carrier that picks up tooling from tool movers and feeds the tool racks in front of the five machining centers in each FMS line

- Tool buffer racks (magazines) that store tools readily ahead of need in front of each machine
- A robotic tool changer to load and unload tools from the buffer rack to the rotary magazine of the machine center

This tool delivery system, one of the best of the 1990s, is extremely responsive at high-paced production. The developer, Fritz Werner of Berlin, Germany, has mastered the complexity very well. The system has been displayed at the Chicago Tool Show and works well at Caterpillar's Joliet, Illinois, plant.

Automated tool delivery software is complex, due to the high number of transactions that are involved. If one part needs 10 tools, and the group of FMSs can run 500 parts, a total of 5,000 tools would be needed in inventory. Following are rules to simplify tool delivery software:

- Standardize tooling and design, so one tool can make several parts.
- Minimize tool changes by processing parts that use common tools on the same machine one after the other.
- Keep high-usage, multi-part tooling resident in the tool magazine of the machining center.
- Minimize tool migration by using it only for one-of-a-kind tools for specific part numbers.

The more rules created, the more complicated the software becomes. Even with these four simple rules, tool delivery software is complex. But the difficulty of developing tool delivery software is worthwhile, particularly on large projects where several FMSs are grouped together. Tool delivery software allows the scheduling computer to direct parts to the next available machine, regardless of the tooling that is currently on the magazine. This random access to machining centers allows high utilization of each machine and a better overall return on investment for the FMS.

Tool scheduling and setting. Figure 14 shows a comprehensive tool room with a terminal (T1) which schedules tool preparation. The action starts when the trigger is pulled to route new parts to the FMS. T1 displays the tool list for the job and schedules tool preparation according to the internal FMS scheduling.

There are two different devices to set tools. S1 is used for normal tolerances. The tool is set at the zero line using a micrometer screw and a magnified picture of the tool tip. S2 is used for closer tolerances. A micrometer screw is used to set the tool as close as possible to the zero line, but the tool setting machine determines the distance from zero. If, for example, the machine determines that the tool is at 0 + 0.003 of an inch when the tool is delivered to the machine, it will automatically be positioned at −0.003 of an inch to adjust for the difference. This requires that each tool be identified and have the capability of carrying the message with it. It also requires that the machine program position the spindle to compensate for the tool setting.

Tool compensation. The zero line of the machine can also be altered to adjust for tool wear. This is called tool compensation. This alteration is done as the CMM shows a deviation or a drift within the tolerance given for a dimension of a part. Tool change is generated when the drift becomes critical and approaches the tolerance limit. Tool compensation should only be done when tolerances are very difficult to hold. Under normal circumstances, this practice should be avoided. It is important to know when tool wear is taking place to maintain a stable process. If you are charting tolerance deviation and tool wear is being compensated for, there will be no evidence that the tool is wearing. It is better to keep process stability by changing tools more often, rather than working with dull tools that are being compensated. Tool compensation, like tool management, has a quality dimension because it is fundamental to stabilize process control.

Tooling technology. A booklet called "Tool Management in Linked Systems," by Helmut Maschke of Scharmann GmbH & Co., Germany, outlines two major reasons why vendors are focused on improving tooling technology.

- Tooling costs can be as high as 50% of total machine costs. To maximize the return on a tooling investment, buyers demand high tool utilization, less tool inventory, and fast tool migration.
- Buyers also need automated tool delivery to maximize the return on the total machine investment. It is automated tool delivery that permits parts to have random access to machinery. This increases machine utilization and improves return on investment.

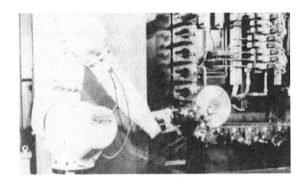

Figure 15 Tool loading manipulators.

Figure 16 Cassette for tool delivery.

There are many examples of good tool management and delivery technology available today. For tool management at the machine, most vendors offer a manipulator that loads and unloads cutting tools from the machine rotary magazine (Figure 15). Some of these manipulators are very light and cause problems. Another option is a gantry robot offered by Scharmann, Pegard, Giddings & Lewis, Ingersoll, and others that serves tools from the machine center buffer racks. Huller Hille has an efficient gantry robot that stores and dispatches a high volume of tooling from a large integrated rack at the machine center.

The trend in tool delivery technology is automated delivery from a central tool room to a machine tool rack. Fritz Werner makes the superb system shown in Figure 14 in which tools are delivered to the FMS by shuttle. A system from Scharmann uses AGVs to present tool cassettes to the machine manipulator (Figure 16). This system is in use at Brown Bowery Corporation in Germany (Figure 17), where it delivers tools from a central tool room to all machinery in one building. Another innovator is British Aerospace near London which has developed advanced tool delivery software, perhaps the best available on the market today. This system,

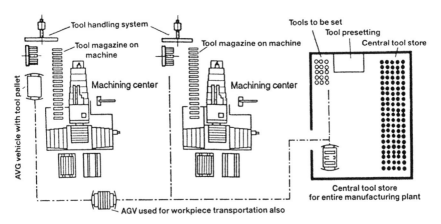

Figure 17 Tool delivery from central room.

which is covered in Chapter 7, features the unique ability to reuse tool holders for different tools, allowing disassembly, assembly, and just-in-time delivery of tools to all machinery in the building.

D. A Full-Scope Introduction

The group of linked FMSs shown in Figure 13 constitutes a full-scope introduction of machining, material handling, logistics, quality control, and tool management technology. Many companies are reluctant to launch an introduction of this magnitude, preferring a longer-term phase-in of new technology. Although a phased introduction of an FMS can help maintain production and minimize startup problems, there are situations in which a full-scope introduction is necessary.

We have seen cases in which an FMS was installed in Phase I, with automated material handling scheduled for Phase II, and an automated coordinate measuring machine (for quality checks) planned for Phase III. But when the FMS began to cut iron, it worked so quickly that it could not be fed fast enough with manual material handling. In addition, it machined parts faster than people could check them for quality. It needed computer-controlled material handling and measurement systems to work at peak production. When implementing a large-scale, high-technology FMS, the best way to get maximum production and a high return on investment is to implement all aspects of the system—including machining, material handling, logistics, quality measurement, tool delivery, and tool management—at the same time.

V.   FLEXIBLE ASSEMBLY AND WELDING SYSTEMS

A flexible assembly system is a manufacturing method that allows products to be built quickly, efficiently, and in any order. It incorporates most of the same principles of flexible machining—the pull trigger, just-in-time delivery, computer controlled logistics, minimum inventory, and integrated quality management. When these principles are applied to the welding operation, it is called a flexible welding system. There are many similarities between flexible assembly and welding systems because a weldment is an assembly of many parts.

A.   Stationary Assembly

Flexible assembly systems feature stationary or stall-built assembly. Stationary assembly is necessary because an operator can stop production at any time for quality reasons. This would not be possible with a moving assembly line. When an assembly line moves at a predetermined, high pace, there will be times when employees are forced to release unsatisfactory quality, just to keep the line going. Stationary assembly gives employees time to complete their work in the highest quality manner.

Figure 18 shows the stationary assembly operation in General Electric's Louisville dishwasher factory. One main overhead conveyor is used to carry only good quality assemblies. An assembly arrives by conveyor at a work station where an operator disconnects it and performs a job. When

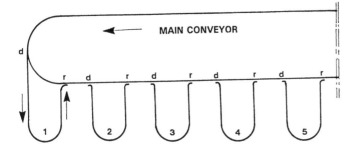

Figure 18 Stationary assembly at GE plant.

the work has been performed to the operator's satisfaction, he or she reconnects the assembly to the conveyor and sends it to the next operation. GE has confirmed that this method generates better quality than the moving conveyor line. GE's overhead conveyor is just one method of facilitating stationary assembly. Kubota in Japan, General Motors-Opel in Germany, and Volvo in Sweden use AGVs to carry products from one assembly area to the next. In these factories, the products remain on the AGVs during the entire assembly process.

Other companies, such as Caterpillar Inc., choose to transport in-process assemblies by AGV, but remove the assemblies at the work stations. The AGVs run in a continuous loop, picking up and dropping off assemblies at the appropriate work areas. In Caterpillar's Decatur, Illinois plant, the world's largest AGV carries off-highway trucks and construction equipment to stations along the "assembly highway" (Figure 19). Because the plant is designed for stall building, each work station has been customized for the operation performed there. Work areas are equipped with the right tools and feature hydraulically powered lifting devices that elevate the product, giving the assembler better access to the under side. This arrangement has generated a quantum jump in quality and productivity compared to what was achieved in a moving assembly line where workers had to climb on the product to do their jobs.

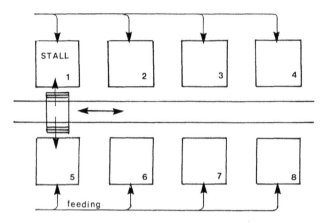

Figure 19 Assembly highway.

B. Split Assembly Lines for Integrated Testing and Adjustment

Although the basic premise of the flexible assembly system is to allow only quality assemblies to move from one stall to the next, there will be situations in which problems arise and adjustments must be made. For this reason, a flexible assembly system includes a split assembly line with an integrated testing and adjustment function. Major assembly work is completed in stalls, then the line splits to allow in-line testing and a lateral exit for products that require rework or adjustments. After being fixed outside the main assembly line, products re-enter the flow and move to final assembly.

C. Kitting

Another key element of flexible assembly is kitting. In a kitting operation, all the materials and hardware needed to perform a specific operation are delivered in a box (or kit) just-in-time to the assembly station. Everything the assembler needs to do his job is in the kit—no more, no less. In traditional assembly operations, workers can spend 40 to 60% of their time walking around gathering material and hardware.

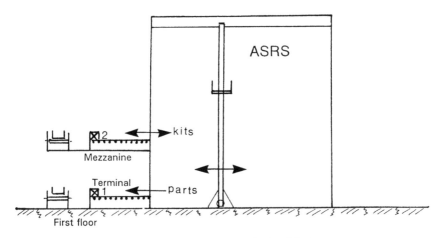

Figure 20   Assembly sequencer, upgrade for kitting.

But kitting can reduce this time to less than 5%. Kitting also saves manufacturing space because there are no boxes of parts and hardware in the work area—just one kit for one job.

Kits are typically assembled one day ahead of need, often on the second floor of the sequencer (Figure 20). Operators working in the sequencer check a computer terminal to determine what should be included in each kit. They retrieve the parts from the sequencer, prepare the kits, and send them to another sequencer for storage. The next day, kits are dispatched from the kit sequencer and arrive on time at the right stall. Assemblers do not wait for kits because they move from stall to stall. For example, if an assembly operation includes 50 stalls, there are 25 teams of assemblers working in 25 stalls. The other 25 stalls are loaded with products and kits waiting for assembly.

D. Flexible Assembly System Layout

Figure 21 shows a generic layout of a flexible assembly system.

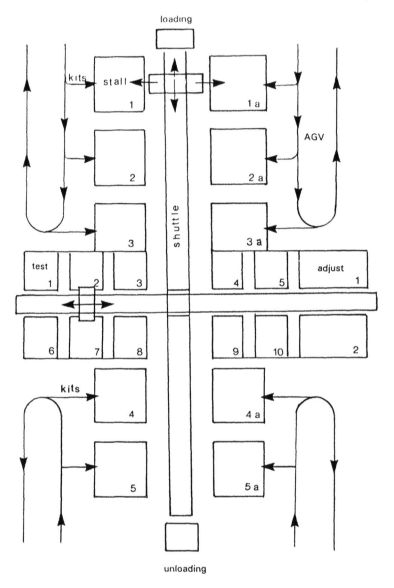

Figure 21   Flexible assembly system (FAS).

- There are 10 assembly stalls, linked by a computer-controlled shuttle.
- Teams of assemblers work in five stalls, while AGVs bring kits to the other unattended stalls.
- The line is split after stalls 3 and 3A to permit testing.
- Two areas are provided on the east side of the layout to permit rework, adjustment, and fitting while there is still good access to the inside of the product.
- The last four stalls are devoted to sheet metal covers, attachments, decals, labels, and finishing work.
- A given product uses four or five of the main assembly stalls.
- Stalls are customized to a specific assembly with all the tools needed for the job.
- People are specialists in one type of assembly for an entire line of product, so they move from stall 1 to 1A, 2 to 2A, etc., during loading time.
- Subassemblies are not shown in the layout, but could be built in stalls with kit feeding and AGV transportation.

Flexible assembly systems produce good results. They are flexible like an FMS because they accept a variety of product options or attachments in any sequence. They are systems because an area computer monitors the delivery of kits by AGV just-in-time to the right stall. Product movement is controlled by operators who use a push-button to activate the shuttle. The pace is computerized with a countdown light (green, yellow, or red).

There is a certain beauty surrounding the activation of a flexible assembly system. The system runs smoothly, efficiently, and quietly. Few, if any parts are on the floor—only kits containing what is needed. There are no shortages, no line runners tracking down missing work. The product flows with a sort of easiness and is built with zero defects as it is fully tested and adjusted before completion. Experience at Caterpillar's Grenoble, France, plant has shown that given comparable production levels, a flexible assembly system reduces direct and indirect labor content by 50% over a traditional assembly operation.

E. Flexible Welding System

The assembly principles illustrated in Figure 19 can be applied to welding. The logistics are the same as in assembly. As parts are needed in assembly, the area computer schedules welding production. All the materials needed for a weld assembly are retrieved from the press shop sequencer, assembled into kits, and carried by computer-controlled AGV to stall 5 where tacking (or spot welding) is done. The fixture which holds the finished product can be carried by the shuttle after tacking is complete. The welder pushes the finish button and the shuttle moves to stall 2 for robotic welding, then to stall 3 for manual finish welding. Inspection takes place at stall 4.

Flexible welding systems are very popular in Japan, particularly at Mitsubishi and Komatsu. Caterpillar uses the concept in Belgium, France, and the United States. Companies using flexible welding report that the process controls the iron supply, helps workers pace themselves, and eliminates shortages and iron chasing.

VI. SUMMARY

Following are key points to remember about automation for the continuous flow factory:

- There must be a balance between flexible automation, logistics, and systems.
- Old machinery can be reused by rearranging it into simple cells which could include subassembly. Simple cells take much floor space and have low utilization, but they fit into a continuous flow factory and moti-

vate the labor force. They should be integrated with computerized material handling.

- Dedicated transfer lines become obsolete when part design changes. Selecting between dedicated (preferably convertible) and flexible transfer lines depends on quantity and variation of part design.
- Computer driven cells are efficient and simple. They have their own justification and can represent the first step of a phased introduction to an FMS.
- Compared to cellular system, the full-scale FMS produces better return on investment, due to better machine utilization.
- A simple merry-go-round FMS is efficient for a large quantity of parts with slight variation. A high volume of iron can be cut by having three simple FMSs, one each for small, medium, and large parts.
- For large families of similar parts, higher technology FMSs are preferred over cell systems. An FMS is easier to coordinate and offers higher utilization and a better return on investment.
- A higher technology FMS is complex and requires a full scope introduction, including an automated parts logistics network, an advanced internal scheduling system, integrated washing and quality measurement systems, and an automated tool logistics network.
- Tool delivery can be extended to an entire building to reduce inventory.
- Quality stabilization for an FMS involves a complete scope of technology, including tool compensation, tool replacement, automated tool delivery, tool room scheduling, and CMM interface for probing.
- Selecting the best type of automation depends on three factors: the quantity of identical parts to be processed as part of a family or lot size; the quantity of like parts within the family or lot variation; and the expected return on investment (which depends on machine utilization).

- Whenever possible, simple technology should be selected. Complex technology should only be chosen if a simple solution is not feasible.

VII. FUNDAMENTAL PRINCIPLES

F28. To utilize older machinery, develop simple manufacturing cells without computer control and link them to the flow by computer controlled material handling.

F29. For new equipment, high utilization is key for good return on investment. Computer controlled FMSs and cellular systems have a better return than simple cells.

F30. There are definite criteria to select between a cell, cellular system, FMS, or transfer line. The key issue is quantity: quantity of identical parts (lot size) and quantity of like parts (lot variation). A return-on-investment analysis must be done to compare the economics of each alternative.

F31. Some applications require a group of linked high technology FMSs. The FMSs must be supported by a full complement of technology to control parts logistics, internal scheduling, quality measurement, and tool logistics.

F32. Stationary stall assembly is the preferred technique for a continuous flow factory. Stalls can be fed by AGV and should be computer controlled.

F33. Kitting and AGV feeding of assembly stalls permit efficient logistics in a continuous flow factory.

F34. Flexible welding systems use the same principles as flexible assembly systems.

5

Overcoming the Challenges
of Automation

I. TYPICAL CHALLENGES AND SOLUTIONS

This chapter presents six typical challenges that may arise when selecting equipment or designing a layout for a continuous flow factory. The first five challenges can be solved with process innovations. The last one can be overcome by applying sound layout engineering principles.

Challenge	Solution
1. Carburizing parts in a central heat treating furnace creates a bottleneck in the flow.	1. Process innovation: the flexible carburizer.
2. Single-spindle machining centers are too slow. More productive multi-spindle heads are needed.	2. Process innovation: the head changer.
3. Machines that have both vertical and horizontal spindles are needed to process the top and lateral faces of a part.	3. Process innovation: the "over" concept.
4. An FMS may not accommodate large parts.	4. Process innovation: the gantry machine.
5. To improve return on investment and respond to assembly needs, gantry machines must accept parts in the order in which they arrive.	5. Process innovation: random-access gantry FMS.
6. When designing a factory layout, there are many obstacles such as services and utilities that disrupt the flow.	6. Layout engineering principles.

II. PROCESS INNOVATIONS

Five typical challenges of automation can be overcome by taking advantage of recent process innovations that include:

- The flexible carburizer
- The head changer
- The "over" concept
- The gantry machine
- The random-access gantry FMS

A. Flexible Carburizer

Traditionally, a heat treat furnace processes parts in batches. A batch is made up of several loads of parts, all with similar heat treat cycles. A sequencer placed in front of the heat treat area can group parts with similar cycles and route them to a furnace where they will be processed together. A short flow interruption is possible only when the number of different heat treat cycles has been reduced through standardization.

One of the most difficult aspects of heat treat to standardize is carburizing. Carburizing is the process that transfers carbon from the furnace atmosphere to the steel crystalline structure. This transfer happens at the surface of the part and extends into the part's structure. A part must be carburized to a certain "case depth," which is the depth that will give the part a surface hardness of 62 on the Rockwell C Hardness Scale after quenching. This degree of hardness makes the part resistant to wear. Case depth depends on temperature and duration of exposure. Different parts require different case depths. The bigger the pitch of the gear, for example, the more case depth is needed. Correction to the shape of gear teeth also affects case depth. Generally, you are heat treating parts with many different case depths, so it takes a long time for the sequencer to accumulate a large enough batch to be processed economically. The longer the sequencer has to wait for iron, the more the flow is disrupted. This problem can be resolved with a flexible carburizer such as the Holcroft Rotocarb™ system, shown in Figure 22, made by Thermo Process System, Inc., Livonia, Michigan.

Holcroft did not invent the flexible rotary carburizing furnace. We saw one at Volvo in Sweden in the early 1980s. What Holcroft did was create a fully flexible, three-stage carburizing system. We believe this outstanding piece of engineering, nicknamed the "triple donut rotary carb," is the only one of its kind. It was developed with input from leading heat treat experts to assure that the system meets the highest level of customer expectations. The system works like this:

- The central heat treat sequencer receives a pull trigger signal from the assembly sequencer to replenish a certain type of gear.

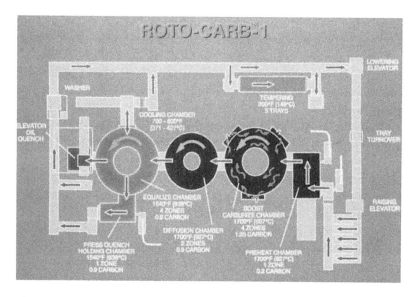

Figure 22 ROTOCARB™.

- The Rotocarb™ computer includes 300 different heat treat cycles. The software makes a "go/no go" decision on every part that is called by the pull signal. The "go" iron enters the system, while the "no go" parts are delayed for the next run because they present too much variation with the other iron.
- Parts that are accepted into the system automatically receive the required heat treat cycle which is initiated as the tray moves through the system.
- Processing begins in the preheat chamber where parts are heated to 927°C. From there, a main pusher moves the iron to the first rotary chamber which is called the "boost."
- The boost is computer controlled and creates the carburizing. It rotates constantly except when loading or unloading trays. The boost has different zones of temperature. Each tray spends the right amount of

time in the right temperature zone to complete the cycle. When the process is complete, the tray is pushed to the next chamber, where the diffusion cycle takes place. Maintaining a leak-proof passageway between the boost and diffusion chambers is the most critical part of this design. The pusher incorporates an oil seal management system to prevent leaks and oxidation.

- In the diffusion chamber, the iron maintains a high carburizing temperature, creating a good transition of carbon content with the core of the gear. After quenching, this gradual carbon content will generate the gradual hardness from 62 Rockwell on the surface to 40 at the core. This makes a good gear.
- From the diffusion chamber, the tray moves into the equalizer chamber. Again, a leak-proof passageway is critical. Equalizing assures that all sections of the part reach the same temperature before quenching. Different cycles are possible in the equalizer chamber. Parts can exit quickly or cool slowly.

The triple donut flexible carburizing concept is complex, but the technology also exists in simple or double donut versions. The triple donut should be used only when sophisticated heat treat cycles are required.

When evaluating the economics of a triple donut, it is important to look at the total return on investment for the factory—not just the return on the furnace. If the furnace is viewed in isolation, traditional technology will show a better return on investment than new technology. But the old technology also takes a lot of space and creates a large, expensive parking lot of iron in the factory as parts must wait to be scheduled. The flexible carburizer improves flow, which will generate a higher total return on investment. We have seen cases in which the new furnace alone produced a return of about 10%, but it helped achieve an overall return on the factory investment of 28%.

B. Head Changer

A head is a heavy-duty, multi-spindle device that is attached to the front of a single-spindle machining center. The single spindle activates the head, which in turn, delivers power and rotation to the other spindles. This transforms the single-spindle machine into a more productive, multi-spindle machining center. Each machining center needs several heads in order to drill, tap, or bore different faces of a part. The challenge is to have the capability of changing heads quickly.

Machine tool vendors in the United States have been slow to develop head changer technology because, as discussed in Chapter 4, most U.S. customers have preferred dedicated transfer lines over flexible transfer lines and FMSs. But as product designs change with increasing frequency, dedicated transfer lines are becoming less feasible in many applications. U.S. suppliers need to improve machine tool flexibility by becoming more proactive in the development of head changer technology.

Head changer technology from Mandelli Industriale Spa, Piacenza, Italy, is shown in Figures 23 through 25. It is being used at the Ferrari car factory which consists of two FMSs and a few cells. Figure 23 is a photograph of the heavy-duty, rotary-type head magazine. Figure 24 shows a device that retrieves the head from the magazine and slides it into the machine. And pictured in Figure 25 is the device that rotates the head so it can be clamped to the spindle. Figure 26 shows a layout of a four-machining center FMS that incorporates two machines equipped with head changers. This FMS has the productivity of a transfer line, but the flexibility to process larger families of parts. Another type of head changer is shown in Figure 27. This is the Orbiter® from Huller Hille GmbH, Ludwigbrug, Germany. The Orbiter is a rotating arm that can remove one head and move a new one into position in 10 seconds.

As Figure 27 shows, this head changer works in four phases. In Phase 1, the multiple drill head on the left has

Figure 23 Head changer magazine.

Figure 24 Head changer slide.

Figure 25   Head changer loading.

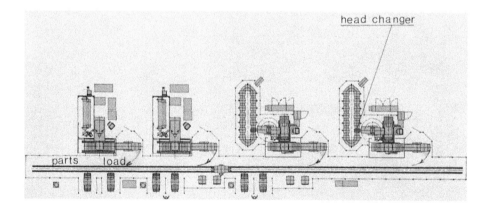

Figure 26   FMS with head changer.

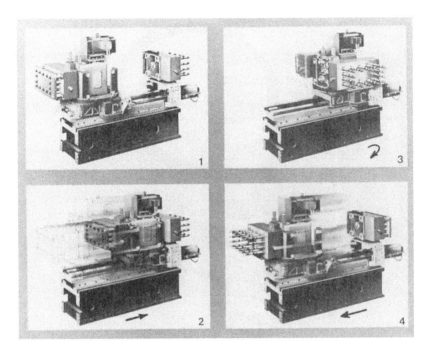

Figure 27 Orbiter®(Huller Hille GmbH).

just drilled 14 holes. The next operation will be performed by the multiple tap head on the right which is approaching from the head magazine. In Phase 2, the tap head is clamped. A fast 180° rotation in Phase 3 brings the tap head to the front end position. The rotation continues in Phase 4 when the tap head moves into the ready-to-work position. At the same time, the drill head in the rear is unclamped and removed, interfacing with a head magazine conveyor or other material handling device. In addition to being fast, the Orbiter is also one of the most rigid and precise head changers available today. One of the best designs of the 1990s, this changer permits many kinds of multi-spindle operations and has paved the way for flexible transfer lines for engines and other products that require highly accurate machining.

Figure 28 "Over" concept.

These two types of head changer technology permit an FMS to run as well as a transfer line with multi-spindle capability, rigidity, and high productivity.

C. The "Over" Concept

Most machining centers are designed to process parts horizontally. A horizontal machine has access to four faces of the part when the table is rotated. A vertical machine has access to one face, the top. Combining both types of machines allows access to five faces. If engineering leaves the sixth face unfinished, the part could be completed in one setup. This would greatly improve factory flow.

Mandelli Industriale Spa has done just that with a patented design shown in Figure 28. Called the "over" concept, this design allows machining in vertical and horizontal positions with rigidity as good as in a traditional spindle. The rotation is programmed on a 45° surface with opposed crowns with Hirth frontal teeth. This guarantees accuracy and the ability to repeat operations. Preloaded Belleville washers supply the clamping force. Tool changes are made in the horizontal position. The 90° spindle rotation is hydraulic. Scharmann of Germany has a similar design with vertical and horizontal capabilities.

The over concept expands FMS flexibility and represents a key technology for the future. Although it is not yet popular in the United States, there is a definite trend toward this concept in other areas of the world.

D. Gantry Machines

A gantry machine (Figure 29) is an effective tool for machining large castings or fabrications. Gantry machines have been used in the aircraft industry for many years, but are not widely used in other U.S. manufacturing operations. Small gantry machines are popular in Europe and Japan. A gantry machine has a slower cycle than an FMS. It offers excellent rigidity for large parts and permits many holes to be drilled, tapped, and bored in one set up. Gantry machines are computer controlled. They are very sophisticated with one head on each side and two heads on the top. Ingersoll of Rockford, Illinois, has excellent gantry machine design. SNK of Japan makes good small gantry machines but does not offer the worldwide design versatility of Ingersoll.

E. Random-Access Gantry FMS

In applications where medium-size parts with more variation will be machined in relatively small quantities, a gantry FMS can be a cost effective solution. With the gantry FMS, parts have random access to machining centers. This improves utilization, reduces the number of machines that

Figure 29 Gantry machine.

must be purchased, and increases return on investment. Waldrich, a subsidiary of Ingersoll, makes a gantry FMS.

We have seen a successful application of a gantry FMS. In this situation, four families of large fabrications were being machined. The company had three options to automate the process. One was purchasing stand-alone machines deployed into four cells. This would have required much space, labor, and capital. The second option was purchasing four gantry machines. This was not cost effective because of low utilization and low return on investment. A third alternative (Figure 30) was to buy three gantry machines and link them together in an FMS. By purchasing three machines instead of four, there was enough capital saved to buy a parts shuttle, a loading station, and a tool delivery shuttle. Adding automated tool delivery to the gantry FMS permitted ran-

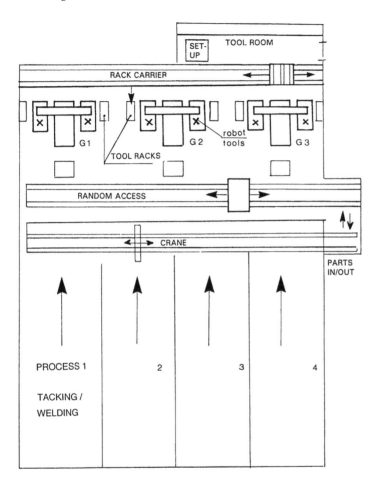

Figure 30 Gantry FMS.

dom access to the three machines. This increased utilization and improved the return on investment. As an additional benefit, this entire FMS is run by only two people—the operator and the tool maker. This is a good success story, and shows what can be accomplished when the factory and machine tool supplier have a strong partnership.

III. LAYOUT ENGINEERING PRINCIPLES

A. Identifying Obstacles to Flow

Figure 31 shows a layout for a perfectly rectangular building. All processes flow from the top of the diagram to the bottom and there are no impediments to flow. The real world is not this simple. As a layout is developed for a continuous flow factory, engineers will discover many obstacles that compromise or stop the flow. Some of these are:

* Cribs for blueprints, gauges, and cutting tools
* Tool room
* Maintenance areas
* Coordinated measuring machine
* Gear lab
* Utilities and transformers
* Shop computer rooms
* Air compressors
* Chip systems
* Factory offices
* Cafeteria and rest rooms

Other obstacles include manufacturing activities that have been centralized for economic reasons, such as:

* Receiving
* Chemical processing
* Testing
* Painting
* Shot blasting
* Heat treating

Some obstacles, such as cribs, tool rooms, and maintenance areas, are movable. Others, like factory offices, cafeteria, and utilities, are permanent. It takes a very competent engineer to locate movable and permanent road blocks in a manner that minimizes disruption to flow.

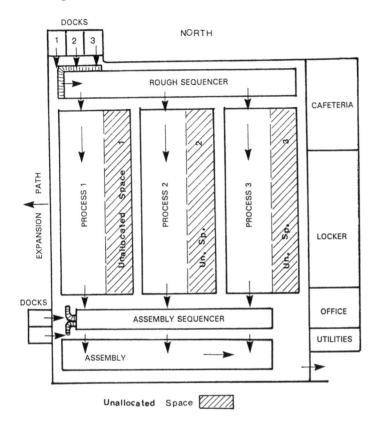

Figure 31 Unallocated space strategy.

B. Space for Growth

A good layout engineer leaves 15 to 20% of total space in each minifactory unallocated. This space is strategically located to permit future capacity increases and migration of new technology without rearrangement of existing equipment. Unallocated space should be concentrated in areas where technological changes or capacity increases are expected. The layout should also include a path for future building expansion. In Figure 31, the path for expansion moves to the west. To accommodate expansion, the west side of the building has no electrical cable, sewage, or utility piping. A concrete slab 100 yards wide could be added to the west side to prepare for the future.

C. Road Blocks

Movable road blocks can be located in unallocated space, then moved if necessary in the future. Permanent road blocks should be located outside the flow. If the flow is from north to south, as shown in Figure 31, permanent obstacles belong in the east or west. In this example, they are located on the east side, in a building adjacent to the factory.

D. Minimizing Energy

In Figure 31, the process carries the iron from north to south. Rough materials are delivered to three docks set up near the rough sequencer. This sequencer dispatches iron east to west to the three processes. Then the processes carry the iron toward the south using integrated material handling devices. Purchased finished goods are delivered to the assembly sequencer docks at the bottom of Figure 31.

There is very little energy expended for the purpose of handling material because deliveries are made to the point of use and the processes carry the iron. Energy consumption is the basis of a simulation which must be done to choose between layout alternatives. The best layout is the one that consumes the least power in 24 hours. The formula for computing energy consumption is:

$$\begin{matrix} \text{weekly} & \text{weekly} & \text{energy} \\ \text{power consumed} \times \text{traffic distance} \div 5 \text{ days} = \text{consumed} \\ \text{(kilograms)} & \text{(meters)} & \text{(kg m/day)} \end{matrix}$$

There is good software available to perform these simulations. Calculations are based on loads and distance. The loads for each layout alternative are the weight of production at capacity. Distance between points depends on the layout. Simulation permits the layout engineer to identify traffic at each critical point of the layout (dock, feeding, process exit, aisle). Heavy traffic spots and bottlenecks can be avoided by changing the flow.

E. Protecting the Flow

Good flow is the root cause of simplicity and productivity in the continuous flow factory. There will always be impediments to flow, but if flow is compromised in too many ways, the full benefits of modernization will not be realized. To achieve the rewards of modernization, there must be a top management commitment to flow. Layout engineers must know that efficient flow is the number one priority. As the layout is designed, every decision made and action taken should enhance, rather then impede, flow.

F. Point-of-Use Delivery

There is no way to design a good layout if all purchased items are delivered to one place. Modern layouts require delivery to the *point of use*. Sequencers manage rough materials and purchased finished parts. There is very little inventory.

One solution for disposing of a central receiving facility is to sell it for conversion to a warehouse. Or, if the facility is adjacent to the factory, it can be converted into an intelligent sequencer for rough or purchased finished materials.

G. Yard Storage and Shot Blasting

Yard storage should be avoided for two reasons: parts rust outdoors, and they must be manually managed. An automated indoor storage system is preferred in a continuous flow factory.

Shot blasting should also be avoided to preserve flow and save supplier packaging. It is better to buy shot blasted parts that are usable when received, then store them under roof, well secured, in known-quantity packaging. Shot blasted plates should also be purchased and stored inside. If it is necessary to have outside storage and a shot blasting facility, plate turnover must be high to avoid rust, appearance, and welding problems.

H.   Chemical Processing

Chrome plating and other chemical processing facilities are expensive and should not be decentralized. But to preserve flow, these facilities should be located near the commodity that uses them most frequently. Other commodities will have to be transported to and from these operations by computer controlled material handling.

I.   Prime Coating Parts

Applying two coats of paint to a finished product is not always best for cost and quality. It is often better to eliminate the prime coat and buy castings and forgings that have already received a rough coat of paint. This job could be done by the supplier, a subcontractor, or a third party logistics company. It could even be done internally in a converted centralized receiving building. If done internally, avoid adding too much lead time and keep the integrity of the rough sequencer location and quantities.

When the prime coat is avoided, final paint costs are reduced. Quality is also improved because the product is protected in places the final paint coat does not access. Hardware can be protected with a plating process. It is costly, but guarantees good appearance and long life. Plating is usually more cost-effective than a prime coat or touch-up on visible nut and bolt heads. There may be situations where prime coating is necessary. The key is to analyze parts thoroughly to determine the best approach.

J.   Heat Treating

Heat treating is a major challenge and represents the most controversial subject in a continuous flow layout. Some heat treat processes such as induction hardening and tempering can be easily decentralized and laid out in a minifactory.

These processes are small and compact, with short cycle times. The utilities to support them can be installed in a machine tool environment, so decentralization does not affect utilization. We found an innovative U.S. machine tool builder who has integrated induction hardening within a machining center, using machine coolant for perishable tooling as a coolant for quenching.

Other heat treat processes like hardening quench and temper or carburizing quench and temper need more construction engineering around them. They also need permanent installations such as generators, coolant, and pumping stations. In these situations, centralized heat treat is the best option. This calls for a compromise, shown in Figure 32, which protects the flow. Here, heat treat has been centralized and is common to four minifactories. A heat treat shuttle carries the iron to a heat treat sequencer. The sequencer segregates the iron by process, depending on the metallurgical specifications. After heat treat, the sequencer directs the iron to the next step in the process (grinding).

Flow in the heat treat area is regulated by an area computer. The iron is pulled from the assembly sequencer. This creates demand at the heat treat sequencer which replenishes itself by calling on processes 1, 2, 3, and 4 to produce blanks. Rough forgings are pulled from suppliers by the rough sequencer.

K. Mezzanines

One of the best locations for a factory area computer room is a fabricated mezzanine. This provides an opportunity for minifactory workers and managers to monitor production status. There should be one area computer per minifactory. It should be located above a place where there are no cranes such as a tool setting area, conveyor, loading and unloading station, crib, or rest room. Factory electrical equipment with microprocessors can also be located on the mezzanine. It should be enclosed in large cabinets with air filtration or air conditioning.

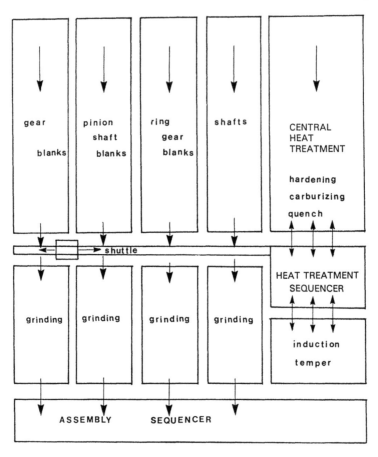

Figure 32 Central heat treat sequencer.

Figure 33 shows a good layout for an FMS with a mezzanine. Machine centers are located on the floor with a parts
shuttle between them. Behind the machines, there are two
shuttles for parts and tool delivery. They transport nonresi-
dent tooling to the machine magazine. The computer room is
on top on a mezzanine with a passageway for visitors. Electrical cabinets are also on the mezzanine. There is a stairway
on the right for access. Under the mezzanine, there is room

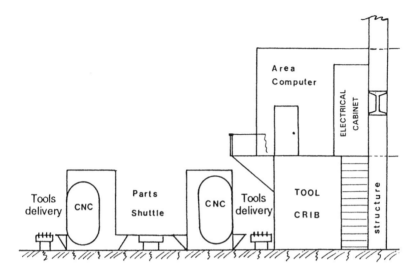

Figure 33   Mezzanine.

for a tool setting area and a crib to store tooling in racks that are ready to be transported by the tool shuttle. Along the length of the FMS, there is room to locate maintenance, rest rooms, or cribs. These areas should be located where posts and columns that are part of the building structure prevent the layout of new equipment.

L. Coordinate Measurement Machines

To avoid material handling and improve response time, CMMs should be laid out at the end of a process or FMS. A shuttle can pick up the iron and route it to the assembly sequencer, or as an option, bring a sample to the CMM and return it to the flow (see Figure 34.)

These layout principles are well known by good engineers. However, we have seen factories where too many obstacles were located in the middle of the flow. This is a costly mistake. Large rectangular areas should be protected without any obstacles or islands, and all activities that are

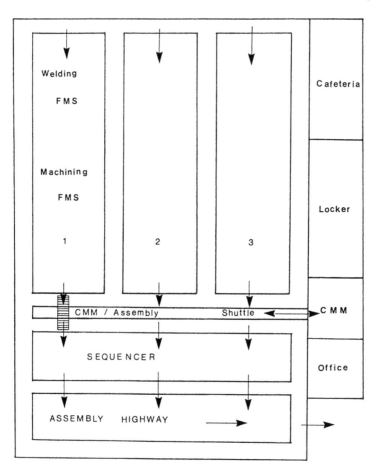

Figure 34 Random access to C.M.M.

not part of the main flow should be located where the posts of the building prevent a large deployment of machinery. Large strips of unallocated space should also be protected from permanent road blocks. They can be used for storage of equipment or material in transit. Sequencers should never be located temporarily unless they are fully manual because they require sizable foundations.

IV. SUMMARY

Process innovations are available to overcome many of the challenges of automation.

- The flexible carburizer improves flow in a centralized heat treat operation.
- A head changer transforms a single spindle machine tool into a more productive, multi-spindle machining center.
- The "over" concept improves utilization by allowing parts to be processed both horizontally and vertically.
- A gantry machine is an effective tool for machining large parts.
- In some applications, a random access gantry FMS is the best choice for high utilization and a good return on investment.

Good layout principles can be applied to protect flow.

- Leave 10 to 15% of total space unallocated for future expansion and technological change.
- Locate unallocated space strategically.
- Put the "movable" obstacles in unallocated space.
- Locate "permanent" obstacles out of the flow.
- Plan for future building expansion.
- Abandon centralized receiving.
- Avoid yard storage and shot blasting.
- Locate chemical processing operations near the commodity that uses them most frequently.
- Buy painted parts and rough materials to avoid priming the final product.
- Decentralize heat treat when possible. When centralized heat treat is necessary, lay it out to support several minifactories and add a sequencer to dispatch iron.
- Build a mezzanine for computer rooms and electrical cabinets.

- Locate a Coordinate measuring machine at the exit of a process.

V. FUNDAMENTAL PRINCIPLES

F35. Road blocks that disrupt the flow can be managed with process innovations and good layout principles.

F36. Carburizing can be done with a flexible carburizer to protect good flow.

F37. A rigid head changer permits the use of multispindle heads in machining centers, a key for flexible transfer and higher productivity FMSs.

F38. Unallocated space should be planned and reserved in strategic places in each process to permit growth and migration of new technology. This will prevent the need for rearrangement in the future.

F39. Layout engineering technology is a key to designing and protecting flow:
- Receive materials at the point of use.
- Rough sequencers are preferred over yard storage.
- Induction heat treat should be decentralized.
- Centralized heat treat should incorporate sequencer feeding and flow back to the process (Figure 32).

F40. Floor space should be protected in the flow. Utilities should be laid out in adjacent areas, not in the flow.

6

Comparing Continuous Flow and Traditional Factories

A continuous flow factory differs from a traditional factory in many ways. The fundamental differences are *type of automation* and *shop layout*.

I. TYPE OF AUTOMATION

A continuous flow factory uses flexible automation, including cellular systems and FMSs. Tooling is resident at the process so there is no setup time and a load of parts can be completely processed from start to finish in the order in which it arrives on the conveyor.

A traditional factory uses stand-alone machines, each of which completes a single operation such as milling, drilling, boring, or tapping. Tooling is not resident. A tool list is generated for each operation, and tools are delivered to the machine, used, then returned to the crib.

II. LAYOUT

The flexible automation used in a continuous flow factory (see Figure 3) is deployed in process blocks or minifactories. All the parts necessary to make a subassembly are produced within a minifactory. That means machining, heat treating, and welding are integrated into each of the minifactories. The minifactory produces a unit of production that is ready to be assembled into the final product. This could be a line of engines, transmissions, pumps, hydraulic cylinders, or hydraulic motors. Or it might be a large part of a vehicle like the chassis and its attached piping and electrical harness. The point is, each minifactory produces an identifiable assembly, and all operations required to make that assembly are deployed within the minifactory.

A traditional factory (Figure 35) is not laid out to produce subassemblies. It includes a machine shop and separate areas (or separate buildings) for heat treating, chemical processing, and welding. The machine shop is generally laid out in lines that produce commodities such as shafts, gears, covers, hydraulics, large cases, and so on. Machine tools are located on both sides of a line. As parts are processed, they often move from one line to another and back again. They also travel to other areas—or other buildings—for operations such as heat treating and chemical processing, then return to the machine shop for finish grinding. It is not unusual for parts to travel for several miles within a traditional factory, following paths like those shown in Figure 35 (one is a casting, the other a weldment).

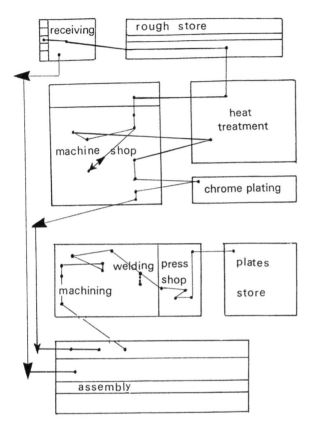

Figure 35   Traditional factory.

III. RUNNING THE SHOP

The fundamental differences in type of automation and layout cause major differences in the way the two types of factories are run.

A. Continuous Flow Factory

In a continuous flow factory, when parts are needed in
assembly, the assembly sequencer retrieves the material,
routes it to assembly, then activates a pull signal to replen-

ish itself. The pull signal tells the material handling system to get a load of rough materials and deliver it to the minifactory where it will be processed. The load is completely processed, then transported to the assembly sequencer. This restores inventory to the minimum level, so the sequencer is ready to dispatch parts to assembly when needed.

In this type of factory, work flow is smooth, continuous, and synchronous. There is very little paper. Production costs are self-contained in the minifactory and easily identified and managed. Manufacturing velocity is high, and in-process inventory is low. There are fewer material shortages, fewer conflicts, and fewer crises. As a result, there is ample time to make quality the number one priority. It takes fewer employees to run the factory, but the work atmosphere is relaxing and jobs are rewarding, so morale and motivation are high.

B. Traditional Factory

Production in a traditional factory begins when a batch of work orders is issued. Each work order lists all the operations that must be done to process one part. A work order also specifies the quantity to be produced, plus starting and completion dates for every operation. A typical work order contains about 10 operations. Each operation is performed by a different person using a different machine. The foreman schedules starting times for every job and procures all the necessary resources. It is not unusual for a foreman to schedule 200 work orders (2,000 jobs) per month, so there are many opportunities for conflicts and problems. All jobs are scheduled separately, and all parts are processed one operation at a time. Parts move through the factory in a stop-and-go manner, stopping after each operation to await scheduling for the next job. When all operations have been completed, the parts are ready to be delivered to assembly.

In this type of factory, work flow is fragmented and interrupted. Large quantities of paper are generated. Cost management is challenging, manufacturing velocity is slow, and in-process inventory is high. The work environment can be chaotic, as much time is spent reacting to problems and resolving conflicting priorities. Shortages are common, and as a result, production quantity is given a higher priority than quality and cost.

IV. A CLOSER LOOK AT THE DIFFERENCES

A. Unit of Production

In a continuous flow factory the unit of production is one load of parts. This is possible because flexible automation permits parts to be fully processed from start to finish.

In a traditional factory, where stand-alone machines are used for each operation, the unit of production is the output of one machine which has performed one of the ten operations listed on the work order.

B. Control Point

Because the basic unit of production is a finished load of parts, a continuous flow factory is controlled at the part number level. The assembly sequencer initiates production of a load whenever a part must be replenished (see replenishment algorithm, Figure 4). This production trigger is random because each part has a different life based on lead time for processing, exact quantity on hand, and real rate of consumption at the assembly line. The trigger signal is random, so only a small amount of "water" (a load of parts) is added to the flow at any point in time. These small additions to the flow permit smoother, more manageable production and eliminate conflicting priorities.

In a traditional factory, the control point is a batch of work orders. Work orders are based on the engineering bill of materials which is the list of all parts composing an assembly. This list gives the quantity of parts needed to build a subassembly. This quantity is multiplied by daily production and by the number of days covered by the work order (for example, 10 or 20 days) to determine the gross requirement quantity. The quantity on hand, which is known from the computer, is used to compute the net requirement quantity. The computer also determines the completion date for the work order and subtracts lead time to set the starting date. Work orders are issued to the foreman in batches. Each new batch creates a wave of production. As the wave begins, the rough machines at the start of the process are very busy, while the finishing machines have little to do. Later, there is too much work at the finishing machines and not enough at the start. This wave effect has a negative influence on productivity and machine utilization. It creates scheduling challenges and conflicting priorities. The precision of computer-controlled scheduling is lacking due to the use of computer file inventories and average consumption data, rather than real quantity used at assembly. This lack of precision causes shortages and emergencies.

C. Conflicting Priorities

The supervisor's job is easier in a continuous flow factory, because there is no conflict of priorities. The area computer causes the flow, scheduling all jobs with sufficient lead time and assuring that parts are processed as needed in assembly. Problems arise only when execution fails due to exceptional circumstances. The supervisor's job is to manage the exceptions when the flow stops due to machine breakdown or human error.

In contrast, it is harder to manage in a traditional factory because there are many conflicting priorities. In any given month, there may be 2,000 or more different jobs competing for resources. As the supervisor schedules all these jobs, there is high potential for errors and shortages on the assembly line. To avoid shortages, many traditional factories use a "short list" system. A one-day short list is developed, specifying all parts (made and purchased finished) that are expected to be short the following day. Then a team of "iron chasers" launches an emergency expediting effort to procure the parts. We have seen plants with up to 50 material control employees writing short lists and expediting material.

Short list writers are often blamed for shortages, so they have learned to develop long, thorough, and costly lists. And in many cases, a one-day short list is not sufficient to prevent shortages, so five-day lists have been developed. A five-day list has far more than five times the number of parts of a one-day list. There is a geometric progression, so if there are 20 items on the one-day list, there may be 300 on the five-day list. A five-day short list creates much confusion in the factory because the large number of emergencies completely disrupts normal scheduling. In many traditional factories, short lists become the major scheduling tool. The schedule indicates that "top priority" parts are those that will be short the following day; "emergency" parts are those that will be short in two days; and "urgent" parts are those that will be short within five days. "Normal" parts, those needed beyond five days, are not even scheduled until they achieve "urgent" (or five-day) status. This is a costly, high-pressure mode of operation and leaves managers little time for quality improvement, human relations, training, or cost reduction activities.

Although the practice of managing by the short list is costly and disruptive, the short-list concept is actually applied in a continuous flow factory. The pull trigger acts as a computerized short list, setting production priorities so that parts will be delivered on time and as needed to assembly. However, the computer-controlled pull trigger is more efficient than an emergency short list because it is based on actual inventory on hand in the sequencer and it gives the internal or external supplier sufficient lead time to produce the parts. As a result, there are no shortages, no emergencies, no waves of production. Parts flow is continuous and synchronous and assembly requirements are met in an orderly and cost effective manner.

D. Acquisition of Resources

In a continuous flow factory, all the resources required to perform a job are captive in the process. Tooling and programs are resident or can migrate automatically, so parts can be processed in any order, and one or two operators can perform all the jobs. This compares to a traditional factory in which a foreman is responsible for acquiring tooling, fixtures, gauges, instruction sheets, and competent operators for every operation. It is a demanding task. To acquire the right tooling, for example, 2,000 or more tool lists must be generated each month. Each must be served by the tool crib, as tools are distributed to the process in time to get the job started, then returned to the crib following the operation.

Acquiring all these resources and resolving the inevitable conflicts that arise demand much management time. As a result, a traditional factory needs more supervisors than a continuous flow factory, and each foreman has less time to devote to quality, cost, and people issues.

E. Quality Management

Supervisors in a continuous flow factory have one key priority—maintaining the flow—which is, by definition, managing quality, because flow stops whenever there is a quality problem.

In contrast, supervisor in a traditional factory are so busy causing the flow and chasing iron that they have little time for quality management. Quantity is usually their number one priority. Specific tools for managing quality in a continuous flow factory include process capability and process control. These disciplines could be used in a traditional factory, but are frequently not used due to lack of time.

F. Lead Time

Thanks to flexible automation, the minifactory layout, and the pull trigger concept, lead time (the time required to process a part from start to finish) is 5 to 7 days in a continuous flow factory, 10 days including assembly.

Lead time in a traditional factory is 20 to 90 days, possibly more, because each work order has an average of 10 operations, all of which must be scheduled and completed one at a time. The obvious consequence of longer lead time is failure to respond to market demands.

G. In-Process Inventory

In-process inventory is very low in a continuous flow factory. The only iron in process is at the cell or FMS. In a traditional factory, in-process inventory is very high, with many tubs of parts stagnating on the floor between operations.

H. Materials Delivery and Storage

In a continuous flow factory, materials are delivered as close as possible to the point of use and stored in intelligent sequencers which are strategically placed to serve nearby activities.

Most traditional factories have separate receiving buildings where materials are delivered. Parts must then be transported to the appropriate locations. This increases material handling costs and material management challenges. Traditional factories also use outside storage of rough parts. Outside storage can cause rusting. It is also more difficult to keep track of materials, because parts rarely have a permanent location outside; people tend to store them wherever there is room. The protection and discipline of intelligent sequencers are preferred for the continuous flow factory.

I. Cost Management

Cost management is easier and more accurate in a continuous flow factory because every minifactory produces a unique, marketable product. All production, support, supervision, purchasing, and material handling costs are self-contained and can be easily identified. And because production is driven by assembly, whatever is produced in the minifactories is demanded by and sold to the customer.

There is very little cost sharing between minifactories, except for heat, light, host computer services, building services, and so on. As a result, costs in a continuous flow factory can be planned, monitored, and controlled with a high degree of accuracy. Most of the equipment is driven by computers with output measurement in line. There is no posting of production. Area computers prepare productivity reports. This compares to a traditional factory where costs are computed only when all the work orders are completed for one assembly. Labor hours used are posted for each operation of a work order. There is much delay between the production date and the date when cost information is available.

V. CONTINUOUS FLOW ADVANTAGES

Based on the differences between the two types of factories, a continuous flow manufacturing operation has the following advantages over a traditional factory.

- *It is easier to manage.* Production is more timely. There are no shortages at assembly, no obstacles to acquiring resources, and no need for complex scheduling systems. Problems are evident when flow stops. Costs can be easily identified and managed.
- *There is more focus on quality and cost.* Managers are not creating flow, chasing iron, and competing for limited resources, so they have more time to concentrate on the vital issues of cost and quality.
- *It is more responsive to customer needs.* Orders can be filled faster because both lead time (between the pull trigger and the completion of the part load) and throughput time (total processing time from order to shipping) are minimized.
- *There is less in-process inventory.* Lead time is short, and loads move through the process with high veloc-

ity. There is no stopping or stagnating between operations.

- *The factory can be productive with fewer employees.* A continuous flow factory requires fewer operators, fewer supervisors, and fewer levels of management than a traditional factory. Some jobs such in material control, quality control, scheduling, factory accounting, and the tool crib can be eliminated all together. Other jobs like material handling can be greatly reduced in number.
- *Employee motivation is higher.* Morale and satisfaction are higher in a continuous flow factory because job content is broad, quality is a high priority, work flow is smooth and even, and the work environment is calm, controlled, and productive.

VI. SUMMARY

The following chart recaps the primary differences between a continuous flow and traditional factory.

Characteristic	Continuous flow	Traditional
Automation	Flexible processes	Stand-alone machine tools
Layout	Compact design made up of minifactories which produce sub-assemblies.	Based on commodities. Long distances between some operations.
Flow	Continuous and synchronous. Caused by area computer. Moves in one direction to feed assembly.	Stop and go. Caused by supervisor. Moves in many directions.
Unit of production	One load of parts.	One operation of a work order.

Characteristic	Continuous flow	Traditional
Control point	One load of parts.	One batch of work orders.
Parts processing	Start to finish.	One operation at a time.
Lot size	Small.	Large.
Set up	None. Process is flexible and tooling is resident or migrates under computer control.	Tool list is generated. Tools are delivered to each operation, used, and returned to the crib.
Material handling	Automated, computerized. Pallet loading is only manual job.	Very physical. Lift trucks, jib cranes, tote boxes.
In-process iron	No iron on the floor. All parts moving by conveyor or AGV to next destination.	Large quantities of iron in tubs or tote boxes on floor between operations.
In-process inventory	Low.	High.
Lead time	5 to 7 days, 10 including assembly.	20 to 90 days to complete a job.
Foreman's role	Maintain flow.	Secure resources, cause flow, chase iron.
Key performance measures	Cost and quality of parts.	Production quantity.
Quality	Operators responsible for quality. No quality control personnel. Flow stops if quality problem arises.	Needs quality police. Flow can continue despite quality problems.
Cost monitoring	Easier and more accurate because most costs can be identified and are self-contained in minifactory.	Difficult and less accurate because most costs are shared and must be assigned.

Production status	Easy to monitor. Visible in subassembly completion.	Hard to monitor. Must look at work order pile and short list.
Problem detection	Visible in bottlenecks to flow.	Not always visible. May go undetected.
Paper	Very little. Data captured by computers.	Paper mill. Data captured manually. Large set of work orders.
Structure	Lean. Three levels.	Pyramid. Five levels.
Unattended operation	Possible for a few hours.	Not possible.
Job content (supervisors and operators)	Broad jobs. Many responsibilities. High sense of accomplishment.	Narrow jobs. Few responsibilities. Little sense of accomplishment.
Accountability (supervisors and operators)	Focused accountability for all aspects of one process (quantity, quality, timeliness, cost).	Shared, diluted accountability.

7

Information Systems for the Continuous Flow Factory

I. FIVE DIMENSIONS OF INFORMATION SYSTEMS

The subject of information systems is deep and complex. To simplify the topic, this chapter will cover five basic dimensions of information systems:

- The *integration* of various computer systems
- The hierarchy of *hardware*
- Factory *software*
- The *network* or physical connection between computers

- The *"islands"* of computer systems that are not connected to the network

II. INTEGRATION

There are scientific definitions of integration, but we choose to define it by describing what it accomplishes. In the continuous flow factory, integration is the linking of computer systems within the factory. This linkage makes it possible for systems to share and create data, generate a continuous and synchronous flow of material from rough to final assembly, and inform management about quality, production output, and cost.

Integration is fundamental to a continuous flow manufacturing. It creates the flow of information that is necessary to drive production. This information flow generates the pull signal from the assembly sequencer and causes the automatic retrieval and delivery of materials to the process. It controls the processing at the cell or FMS, and it manages the automatic delivery of finished parts, just-in-time, to assembly. Integration is key to the success of a continuous flow factory, but it is the most complex and challenging aspect of the vision.

A. Full Integration

Full integration, shown in Figure 36, means that all computer systems in the enterprise—including engineering, marketing, finance, and manufacturing systems—communicate and interact with one another. All functional areas are linked electronically, so work groups can exchange information, update files, and share data. Information flows freely across the entire organization, and virtually all paper has been eliminated. This level of integration is still a dream, and will likely remain so for many years. In fact, no one is certain that the benefits of full integration will ever exceed the costs of achieving it.

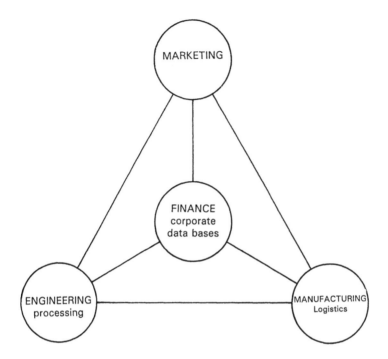

Figure 36 Dream of full integration.

One of the most costly aspects of full integration is developing the data base for the system. A data base is a storehouse of information which can be accessed by users throughout a company. A data base for a fully integrated factory would be extremely large. With today's technology, it is not possible to build up a data base of this magnitude, keep it current, and retrieve information from it in an organized manner.

Today the most widely used data base system is ORACLE™. It will not permit the enormous number of transactions necessary for full integration, but many companies find it useful for smaller scale data bases. IBM is working hard on data base technology and has announced different versions of DB², with more development promised. Until data base technology is significantly enhanced, the paperless factory is still a dream. But in the meantime, a company can take steps to achieve a realistic level of integration now, while preparing for longer term integration in the future.

B. Achieving Realistic Integration

A more realistic level of integration for a continuous flow factory is shown in Figure 37. In this vision, the engineering system, logistics software (MRP II), and the factory floor are linked by an "operation data base." This data base, located in the host computer, is a combination of all existing data bases from the three functional areas.

Figure 38 shows the build-up process for the operation data base. For a period of time, existing data bases (DB1, DB2, DB3, DB4) migrate to the operation data base through a translator (T1). The operation data base supports existing

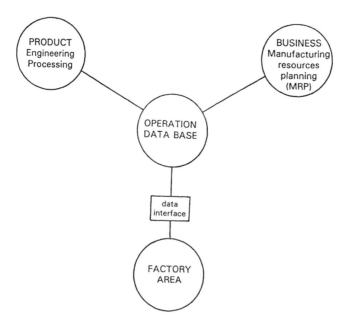

Figure 37 Realistic integration.

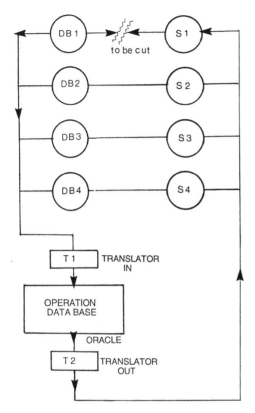

Figure 38   Migration to operation database.

systems (S1, S2, S3, S4) using another translator (T2). The basic philosophy is to keep the factory running using existing systems, while creating a new single data base. ORACLE™ software is effective for this type of integration, because only a portion of the corporation's systems are being linked, so the number of transactions is relatively small. The roles of T1 and T2 are complex and go beyond the scope of this book. But to simplify their functions:

- *T1 is a populator.* It translates information received from existing data bases, putting it in a usable form for the operation data base. It also keeps the operation data base current. T1 takes data from existing data bases and creates an electronic entry into the operation data base. This eliminates human intervention, improving accuracy and reducing the cost of building the data base.

- *T2 is a translator.* It supports existing systems connected with the operation data base. Then as connections between DB1 and S1, DB2 and S2, etc., are progressively eliminated, the operation data base begins to activate the existing systems. Down the road, it will activate new systems, and S1, S2, S3, and S4 will be eliminated.

In order to achieve this level of integration, the company must develop corporate specifications for factory integration. Working with equipment vendors and systems developers, an organization must define the information flow required between the factory operation systems. These specifications discipline the transaction format and permit communication between all factory systems.

C. Preparing for Future Integration

Despite the fact that full integration is a dream today, information technology is changing rapidly, so it makes good business sense to prepare for future integration. Preparing for the future requires that data bases are organized, standardization takes place, and all activities that could jeopardize long-term integration are avoided. Four ways to prepare for the future are:

- *Standardize computer aided design and computer aided manufacturing (CAD/CAM) systems.* Different work groups or business units may demand different CAD/CAM systems, but to maximize integrability (or the ability to be integrated), a company must select a single CAD/CAM system. This may require a directive from top management.

- *Use three-dimensional (3-D) design.* Today most companies design products in two dimensions. However, machines, robots, and the entire manufacturing world are in three dimensions. Three-dimensional product design will permit future integration of engineering and manufacturing, allowing faster and less costly programming of factory machinery and equipment. The aircraft industry pioneered the use of 3-D design. Boeing uses CATIA software, created by Dassault of France and modernized by IBM. McDonnell-Douglas' system, UNIGRAPHIC II, is also a state-of-the-art system. Three-dimensional design is the pathway to the future. A functional organization with separate engineering and manufacturing departments may have trouble migrating to 3-D design because engineering endures the pain, while manufacturing reaps the benefits. Strong corporate leadership and a total business perspective are required to move an organization to 3-D design.
- *Standardize engineering work stations.* Some companies prefer powerful, personal computer–based work stations, believing that decentralized computing is faster and allows engineers to be more productive. Other organizations favor centralized computing, claiming that the slight difference in speed does not affect productivity. It does not really matter whether a company chooses decentralized or centralized computing. What is important is that an organization standardize with a common work station and three-dimensional software.
- *Standardize Manufacturing Resources Planning II (MRP II) software.* If a company has different units producing different components for one product, all units should have the same MRP II software. Choosing a corporate MRP II system is a critical issue. Some organizations buy the software, others elect to develop their own. Today, after years of improvement,

COPIC™ from IBM is a valid system and may save a company the pain of designing its own system. Still, any purchased MRP II requires some changes in the way an organization works. It is generally more cost effective for a company to change practices to adapt to the MRP II, than it is to change the MRP II to adapt to the company's practices. If an organization is willing to change procurement rules, a purchased MRP II may be sufficient. Otherwise, the system must be developed internally. The MRP II decision cannot be delegated. A committee can propose a course of action, but a high-level executive must be involved, and the final decision should stand. Changing an MRP II is costly and highly disruptive and should be avoided.

III. HARDWARE

In broad terms, there are four levels of computers in the hardware hierarchy of a continuous flow factory. Three levels are shown in Figure 39.

- *Level 1* computers are located at the process—in the cells and FMSs.
- *Level 2* hardware includes the area computers—the machines that drive the minifactories.
- *Level 3* is the plant host computer. It houses the operation data base and is the link to *Level 4*, the corporate business system, which is not a part of the continuous flow factory and will not be discussed.

A. Level 1

Most of the time, Level 1 computers for cellular systems or FMSs are furnished by the equipment vendor. They are either made by the machine builder (for example, Giddings & Lewis) or by a computer manufacturer (Digital Equipment Co. is a frequent choice). Some vendors, mostly in Germany, use Siemens.

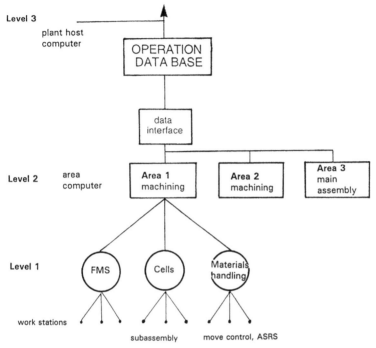

Figure 39 Factory system hierarchy.

Level 1 computers control the process. In an FMS, for example, a Level 1 computer manages virtually all aspects of production—tooling, numerical-control programming, material handling, and quality. An operator is necessary to feed and monitor the FMS—a computer handles everything else.

Figure 40 shows a Scharmann FMS network. This is an advanced FMS because it includes tool delivery, a coordinate measuring machine, and a washing machine, and it integrates information for deburring, assembly, and other operations. The FMS computer is a Micro VAX 32 from Digital Equipment Corporation with a 16-megabyte magnetic tape memory and two 159-megabyte disk drive memories, supplemented by a 296-megabyte magnetic tape memory for data back-up. The FMS has three machine centers, but could

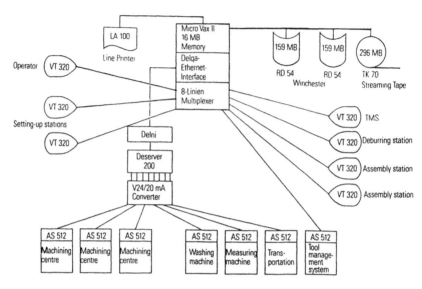

Figure 40 Network for FMS.

have as many as 10. The network is a star configuration so it can accommodate washing, measuring, transporting, and tool management equipment with the V24/20 mA converter from Digital. This system utilizes Ethernet, a Xerox product for Local Area Networks (LANs) that allows units to be linked for storage and processing. The interface to Ethernet is obtained with Delni. The eight-line Multiplexer takes care of all the VT320 terminals.

Connecting the computers, terminals, numerical control machines, and equipment shown in Figure 40 is costly and requires much expertise. Equipment vendors have that expertise. Take advantage of it. Use their experience and knowledge. Do not ask them to make changes to their systems. If you do, you will end up with prototypes, not proven systems. Your only requirements should be that the vendors' systems perform the functions required by the process, conform to corporate specifications for factory integration, and communicate with higher level computers.

B. Level 2

The Level 2 or area computer is the centerpiece of the continuous flow factory vision. It is connected to process computers and sequencers and drives minifactory production. The area computer causes flow. It is the primary data base and computing unit for the minifactory. It provides a realistic level of integration within the minifactory and is workable with today's technology because the amount of data is more manageable at the area level. Only a few companies on the leading edge of factory integration understand the value of area computing, but there is a growing interest in it.

One successful example of area computer integration is located at Caterpillar's Grenoble, France plant. This minifactory, integrated by Voest Alpine of Germany, includes:

- Five stand-alone machines without computerized numerical control (CNC). These machines are about 15 years old (Warner Swarsey–Lorentz Shaper).
- Ten stand-alone machine with CNC which are about 6 years old.
- Two FMSs from different vendors. One is ten years old and the other is new.
- Five cells with cell control computers of different brands.
- Induction heat treat.

In this minifactory, more than 50 machines are linked together by a material handling shuttle. The entire minifactory behaves like a single FMS with loading and unloading stations for the whole area. Operators of old equipment in primitive cells communicate with the computer using push buttons that deliver start and finish signals. The iron is cut in perfect synchronization with assembly demand. There is no buffer inventory. The complex is responsive and coordinated. It requires only one pilot and very few operators. There are fewer supervisors and white collar employees. The inspection and tool room functions have been eliminated. This minifactory has achieved very aggressive cost targets.

Integrating systems at the area level is key to the realization of these targets, as are layout and logistics.

C. Level 3

This is the plant host computer, generally an IBM product. It can exchange data with area computers and is connected to the corporate computer. The trend today is to combine Levels 3 and 4 to get higher utilization of the corporate computing system. The implementation of area computers should accelerate this trend, as the area computer will satisfy plant needs, and the higher level computer will handle the rest.

IV. SOFTWARE

A. Basic Software

A continuous flow factory requires three basic types of software:

- Material handling management systems
- Cell controller and FMS systems
- Area computer systems

Material handling management systems control devices such as AGVs, monorails, gantry robots, intelligent cranes, shuttles, carriers, and others. This software could also control fork truck movement if the trucks are equipped with radio frequency terminals. Some material handling devices are captive to one cell or FMS and are handled by the process computer. Other devices are common to several cells or FMSs and must be scheduled by an area computer. Material handling software manages scheduling, assuring delivery and exit of materials during the right window of time. It also coordinates material movement when more than one handling device is used. Sources of material handling software include Digital, EDS, and other suppliers and integrators. Purchased material handling software must be amended to meet corporate specifications for factory integration.

Cell controller and FMS systems control the functions of the cell or FMS. Typical functions for a cell controller system include:

- Downloading numerical control programs from the area computer and maintaining a program library
- Controlling work sequence according to a list of priorities from the area computer
- Checking resource availability and amending priorities
- Assisting operator by displaying instructions, setup, and resource information
- Providing maintenance resource status information
- Providing management control information such as simulation of output, data collection time, machine status, scrap, rework, inquiries from terminal, and machine performance
- Interfacing with sequencer and material handling management systems

FMS and cellular system software are better when provided by the machine builder. Cell controller software is available from British Aerospace, EDS, and other vendors. Purchased software must be modified to fit corporate specifications. It is normally best to purchase one cell controller package and make it available to the entire company. This will permit future integration.

Area computer software has one primary role: creating flow. It has the following functions:

- Managing sequencer inventory and responding to production commands (Figure 4 algorithm)
- Interfacing with lower level computers: assigning work in sequence to cells, FMSs, and subassembly to ensure completion during a window of time
- Generating move commands for material handling systems based on production sequence
- Monitoring minifactory production and providing status reports and alarm signals as needed

- Interfacing with higher level computer for production reports, cost input, inventory status, and factory management

The relationship between a good minifactory layout and area software is often overlooked. If the factory layout is good, if the parts flow naturally, and if the pull signal is used, area computer software is cost effective. Flow is automatic, and there is little need for scheduling, routing, or coordination. As a result, fewer data transactions are required and less computing power is necessary. This reduces information system costs significantly.

B. Software Principles

Software for a continuous flow factory is based on the following principles.

1. Modular, transportable software

In a large company with several business units, it is common for each unit to hire its own software developer and create its own systems. We have seen large corporations with five CAD/CAM systems and three MRP IIs. This is obviously a costly approach, and results in much duplication of effort. But more importantly, it makes longer term integration difficult, if not impossible.

A better approach is to develop corporate-wide modular software. Whether it is created internally or with help from external consultants, software for the continuous flow factory should be developed at the corporate level with much involvement on the part of business units. The goal should be development of good generic software that can be "transported" throughout the company. An example of modular software is shown in Figure 41. It contains a core component that satisfies the overall vision of the continuous flow factory. It also has a secondary component, represented by the circle, that reflects local enhancement or adaptation. The

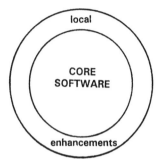

Figure 41 Transportable software module.

core element is provided at no cost to the business unit, while the secondary portion is funded by the unit.

Good modular software produces many benefits. It assures that all systems support the overall vision, yet respects local requirements. It also reduces costs and protects the company's ability to integrate in the future.

2. Control over factory-related software

In order to achieve a realistic level of integration today and make future integration possible, it is vital that control of engineering, manufacturing, and logistics systems be centralized. This requires a strong, competent corporate factory information systems staff, reporting to a high level executive. Traditionally, all information systems staff report to a financial executive. But during a modernization program, it may be more appropriate to have the following factory-related systems managed by the executive accountable for modernization:

- Product design (CAD)
- Process development (CAM)
- Cell, FMS, material movement
- Order entry, product delivery
- Logistics, MRP II, quality, supplier inventory
- Area computing

3. Standards

Centralized control command makes it easier to develop and maintain unified standards. We have already discussed the importance of standardization in a continuous flow factory. Standardization affects all aspects of modernization, but is particularly critical for successful integration. Key areas that must be standardized include:

- CAD/CAM systems including 3-D design software and work stations
- MRP II between units
- MAP standards from vendors
- Corporate specifications for factory integration
- Clear policy on the respective roles of IBM and DEC
- Decision to implement modular, transportable software

4. The role of software

Software plays a vital role in a continuous flow factory. However, it is important to remember that it is just a tool. It exists only to support and serve the vision. Without good flow, an efficient layout, flexible automation, effective logistics, and a clear set of operating principles, software cannot create an efficient and productive factory. Software is a cost effective investment only when it serves a well-designed project.

5. Production at the part-number level

As explained in Chapter 6, a single part number is the basic unit of production and point of control in a continuous flow factory. In contrast, the basic unit of production in a traditional factory is a single machine operation, and the point of control is a batch of work orders needed to complete an assembly.

The systems that support a continuous flow factory vision schedule work at the part number level. This scheduling is based on the algorithm in Figure 4 and the pull-trigger concept. The pull trigger serves an as electronic Kanban, scheduling production as parts are consumed in assembly. The Kanban idea has dominated production control in the Eastern world for decades. The concept of electronic Kanban is a recent innovation in the Western world. It greatly simplifies factory computing systems for two reasons.

First, because the electronic Kanban controls just one part at a time, a smaller quantity of data is needed to initiate material movement, tooling and programming changes, or quality management. Iron enters the flow in small quantities, so information can be generated in small, manageable amounts. As a result, the factory information system can be smaller and less complicated. In a traditional factory, it takes about ten times more data to drive production. There are ten or more starting times for the operations on the work order, ten movements of material, ten-tool lists, and so on. To generate this volume of data, a larger and more complex information system is necessary.

The second reason why the electronic Kanban permits smaller, simpler systems is the random nature of the pull trigger. The area computer generates a pull signal when the exact quantity on hand is low enough to require replenishment (Figure 4 algorithm). The quantity on hand is known with great precision because of bar coding and electronic identification of parts. Parts are consumed at different rates, so the trigger is pulled at random. This not only smoothes the flow of materials, it also streamlines the flow of information.

In contrast, there is no randomness in a traditional factory scheduled by MRP II. Large batches of work orders are issued to make an assembly, and quantities on hand are based on average consumption data and computer inventory files. To compensate for the lack of precise information about quantity on hand, buffer inventory must be produced. Work orders are issues in large batches, generating a wave of production. This wave must be driven by a corresponding surge of information. To accommodate these surges, larger, more complicated systems are needed.

The MRP II, which operates at the product level with a complete bill of materials, is useful for planning rough material requirements, capacity needs, and employment levels. But the MRP II is a forecasting tool, not an execution tool. It is important to remember that most iron will not be scheduled with the MRP II. Nor will it be scheduled with the pull-trigger system. A typical scheduling breakdown for a continuous flow factory is:

Scheduled, managed, and delivered to point of use by suppliers	40%
Controlled by minimum inventory logic (low cost hardware)	20%
Controlled by pull trigger (includes core make and core buy parts)	30%
Scheduled by MRP II (delivered to assembly line or sequencer)	10%
Total	100%

6. Area computers

The use of area computer systems makes integration easier and more affordable. Each minifactory has an area computer system which synchronizes flow for all the part numbers required to produce one subassembly. An area computer system monitors production of about 500 parts. This compares to a total plant system that could control 30,000 or more parts (10 products with 3,000 parts each). By using area computers, the total computing task is divided into smaller, more manageable segments. This obviously reduces costs. It also makes the systems easier for employees and managers to use and permits a high level of integration using today's technology.

7. Shared systems

As a rule, systems that are shared at the building level are complicated and costly. There are some situations in which separate systems are not feasible, but in general, it is wise to do as much computing as possible with separate area systems.

8. Data and programming

Data and programs should be kept at the lowest possible level. Software at the machinery level should include FMS or cell controller programs, tooling programs, and FMS captive material handling systems. Area level software should control the rough material and assembly sequencers, as well as area material handling systems. Minimizing transactions between the process, area, and host computers will permit the lower level systems to work faster, more efficiently, and more economically.

9. Full-scope installation strategy

To achieve the full benefits of modernization, all aspects of the project including machinery, logistics, and systems must be fully implemented. Delaying the integration of logistics and systems can reduce total benefits significantly. When planning a full-scope installation, it is important to include an adequate budget for the cost of integration. Integration costs are often underestimated or not even considered during the early phases of a modernization project. Integration costs typically total at least 5% of the cost of the machinery.

Some companies attempt to reduce integration costs by doing the work themselves. However, this approach is not always more cost effective. Most of the successful plants we visited used external suppliers. Machine vendors are usually capable of doing the integration around a cell or FMS. Integration at the area level has proven to be most effective when done by an experienced integrator such as EDS USA, EDS Europe, Voest Alpine (Germany), and C.J.P. Alsthom

(France). When using outside integrators, it is beneficial to define the boundaries of the project and estimate the costs before work begins. To reduce costs, focus on integrating process-level and area-level computers. Higher level integration with the plant host computer can be delayed without a significant impact on benefits.

V. NETWORK

Network technology goes beyond the scope of this book. However, the subject will be reviewed briefly. The network refers to physical connections that link different computer systems. There are many ways in which systems can be linked. Outside the factory, computers can be connected by modem modules which use telephone lines to transfer data. They can also be linked via satellite, permitting data transmission through space and paving the way for remote central processing.

Inside the continuous flow factory, the network consists of connections between Level 1 (process), Level 2 (area), and Level 3 (host) computers. It is important to identify and locate these connections at the time the minifactory plan is developed. It is also important to budget for the network. In our view, the network linking Level 1, 2, and 3 computers should be included in the total investment needed for a minifactory.

Creating a computer network to link factory computer systems requires that a standard format for moving data be developed. In the last decade, General Motors has introduced the Manufacturing Automation Protocol (MAP) which provides standardization of data transfer and communication between computers. MAP regulates computer communication within a plant. This standard, broadly accepted in all industries, permits the linking of systems made by different vendors using different technology. All vendors must comply with MAP standards, and must also meet corporate specifications for integration to meet the unique needs of a company's systems.

Computer network technology is moving very quickly. Inside the continuous flow factory, we can see boxes, connectors, and translators physically in the form of coaxial cable or fiber optics. The speed and volume of data movement through the network is increasing dramatically and will continue to do so. These connections are called the Local Area Network (LAN). Different brands of LANs such as Decnet, Ethernet, and Hyperchannel serve different niche applications.

VI. COMPUTING ISLANDS

Some computing systems may operate independently from the continuous flow factory network. These "islands" can include:

- *Research computers.* Specially designed for large mathematical calculations, they require their own memories and operate apart from other factory systems.
- *Personal computers (PCs).* PCs have been the key factor affecting the information system explosion of the last ten years. Many factory systems were piloted with PCs running purchased or homemade software. In recent years, there has been a profusion of PC-based factory logistics and quality systems. In the continuous flow factory, simple cells can be run by PC. However, PCs should not be substituted for Level 1 and 2 computers. PCs should be incorporated into the total system plan for the continuous flow factory. The long-term plan should assure that PC data bases are easily accessible in the future and that PCs will be able to be integrated into the total factory system.
- *CAD/CAM systems.* These product and process engineering systems are often partially disconnected from the continuous flow factory network. However, the engineering data base for the CAD system is generally linked to the factory network. The CAM system

has many applications which require integration with the CAD system or the factory network. These applications include layout design, parts processing, tool design, and off-line programming of robots or machine tools.

VII. SUMMARY

Key points from the chapter:

- A continuous flow information system is an integrated system that includes hardware, software, a network that links the computers physically, and islands of independently operated computers.
- Complete integration of all corporate information systems is not possible with today's data base technology and may not be desirable or cost effective.
- A realistic level of integration between product design, logistics, and manufacturing systems can be achieved using current data base technology. The keystone of integration is building an operation data base.
- To prepare for future integration, an organization must standardize work stations, the CAD/CAM system, three-dimensional software, and the MRP II.
- There are three levels of hardware in a continuous flow factory: process computers, area computers, and the plant host.
- There are three basic types of software for the continuous flow factory: material handling systems, process systems (cell or FMS), and area computing systems.
- Software for a continuous flow factory should be modular and transportable, under central control, and supportive of the factory vision. It must schedule production at the part-number level. Shared systems should be avoided and data should be kept at the lowest possible level.

- Achieving the total benefits of modernization requires a full-scope installation of technology and integrated information systems.
- The cost of integration is at least 5% of the cost of the machine and should be budgeted at the time of purchase.

VIII. BUSINESS OPERATING RULES

R9. Standardization is necessary to integrate now and prepare for future integration. Major areas of standardization include the MRP II, CAD/CAM work stations, engineering graphics system (3-D), the MAP protocol, and corporate specifications for integration.

R10. A continuous flow factory information system links engineering, logistics, and the factory floor with a common data base called the "operation data base" which can be built up from existing data bases.

R11. Corporate specifications for factory integration must be developed in order to define the information flow between systems and prepare for future integration of data bases and systems. Vendors must follow the MAP protocol.

R12. A continuous flow factory uses modular software that is transportable between business units.

IX. FUNDAMENTAL PRINCIPLES

F41. There are three levels of computing in a continuous flow factory. Level 1 is at the process (the cell, FMS, or materials handling). Level 2 is the area computer, the centerpiece of the continuous flow factory. Level 3 is the plant host computer.

F42. Software for the continuous flow factory is based on the following principles:
- It is modular and transportable.

- It is under central control.
- Standardization has taken place.
- Systems serve a factory layout that has good flow.
- Scheduling is performed by the pull trigger at the part level. It is an electronic Kanban.
- Shared systems are avoided when possible.
- Data is kept at the lowest possible level.

F43. Basic software in the continuous flow factory includes material handling systems, cell or FMS controller systems, and area computing systems. There are vendors for modular software that performs these functions.

F44. Area computer software must work effectively in a multi-vendor environment. Integrating multiple systems is a specialized task which may require an outside integrator.

F45. Iron in the continuous flow factory should be segregated. An example is:
- 40% controlled by suppliers
- 20% controlled by minimum inventory logic
- 30% controlled by the pull trigger
- 10% controlled by MRP II

F46. An area computer serves a minifactory, creating and synchronizing flow using the pull-trigger concept. It uses the intelligent sequencer algorithm (Figure 4). The area computer concept allows data to stay at the lowest possible level.

F47. A continuous flow factory must have full-scope installation including computer-controlled material handling.

8

Taking Action and Realizing the Gains

In the previous chapters, 12 business operating rules and 47 fundamental principles of continuous flow manufacturing have been presented. This chapter recaps the rules and principles by category, then translates them into actions to be taken and gains to be realized. The numbers presented in this chapter are conceptual. They are based on existing examples, but must be adapted to each company's situation.

I. RULES AND FUNDAMENTAL PRINCIPLES

A. Strategy

R1. Business operating rules define an environment in which a continuous flow factory will achieve maximum gains.

R2. Modernization should be considered in context with other strategic options.

R3. The benefits of modernization can be leveraged by combining it with other strategies such as continuous cost reduction, simplification, emigration, or reorganization into profit centers.

B. Rationalization

R4. A continuous flow factory is designed after conducting a comprehensive market survey, developing a product strategy, and defining a product line for the future.

R5. A continuous flow factory operates with good capacity utilization, but is capable of growing by steps within unallocated space.

R6. A continuous flow factory makes a minimum number of parts based on economies of scale and proprietary design.

C. Purchasing and External Logistics

R7. A continuous flow factory is very closely tied with a small number of competitive suppliers who enjoy a volume advantage and commit to vital practices such as just-in-time delivery, quality and quantity certification, and electronic identification of load.

F22. Suppliers deliver parts in standardized packages which contain known quantities and have bar code identification labels.

F23. Supplier delivery is performed at the point of use in different zones of the plant.

F24. External logistics may be provided by third-party organizations.

F25. Suppliers may have access to assembly locations and replenish bins as needed, being paid by loads without paperwork.

D. Design for Manufacturability

R8. A continuous flow factory uses simultaneous product and process engineering to design for manufacturability. A design for supplier manufacturability program is also in place.

E. Layout

F5. Material flows in one direction.

F38. Unallocated space should be planned and reserved in strategic places in each process to permit growth and migration of new technology. This will prevent the need for rearrangement in the future.

F39. Layout engineering technology is a key to designing and protecting flow:
- Receive materials at the point of use.
- Rough sequencers are preferred over yard storage.
- Induction heat treat should be decentralized.
- Centralized heat treat should incorporate sequencer feeding and flow back to the process (see Figure 32).

F40. Floor space should be protected in the flow. Utilities should be laid out in adjacent areas, not in the flow.

F. Flexible Automation

F6. Loads of parts are started and completely finished within one machine process and are machined in the sequence in which they arrive in the process loading area.

F9. Machines are flexible and do not need set up time to start and complete the loads assigned to them.

F10. Machine programs and tooling are resident or can migrate quickly.

G. Equipment Selection

F1. Select a reasonable level of automation using proven technology.

F2. Use standard machines when possible and avoid pushing vendors beyond their experience and capabilities.

F28. To utilize older machinery, develop simple manufacturing cells without computer control and link them to the flow by computer controlled material handling.

F29. For new equipment, high utilization is key for good return on investment. Computer controlled FMSs and cellular systems have a better return than simple cells.

F30. There are definite criteria to select between a cell, cellular system, FMS, or transfer line. The key issue is quantity: quantity of identical parts (lot size) and quantity of like parts (lot variation). A return-on-investment analysis must be done to compare the economics of each alternative.

F31. Some applications require a group of linked high technology FMSs. The FMSs must be supported by a full complement of technology to control parts logistics, internal scheduling, quality measurement, and tool logistics.

F35. Road blocks that disrupt the flow can be managed with process innovations and good layout principles.

F36. Carburizing can be done with a flexible carburizer to protect good flow.

F37. A rigid head changer permits the use of multi-spindle heads in machining centers, a key for flexible transfer and higher productivity FMSs.

H. Quality Management

F11. Quality management is integrated into the process.

F18. If there is a quality problem in machining or assembly, flow stops.

F19. Machines, equipment, and all processes have the capability to meet design tolerances. Simultaneous engineering has insured that design and process are compatible.

F20. Machinery has process control for production to be within tolerances. Process control can be assisted by manual gauging, automated gauging, or coordinated measuring machines.

F21. Certify internal and external suppliers for good quality.

I. Flexible Assembly and Welding Systems

F32. Stationary stall assembly is the preferred technique for a continuous flow factory. Stalls can be fed by AGV and should be computer controlled.

F33. Kitting and AGV feeding of assembly stalls permits efficient logistics in a continuous flow factory.

F34. Flexible welding systems use the same principles as flexible assembly systems.

J. Internal Logistics Systems

F3. An integrated factory information system is needed to coordinate delivery from suppliers, regulate movement of material, and support the processes by machine program download and tool management.

F4. Regulate the continuous flow factory by computer. Human intervention is necessary to supervise and assist flow, not cause it.

F7. Parts flow by using computer controlled material handling. Management's job is to maintain flow—not cause it.

F8. Flow is synchronous. A load of parts must be started at the right time to be completed just ahead of need in assembly.

F12. Material handling devices maintain the flow. They can be manual, but they are all monitored by a computer that causes and synchronizes the flow.

F13. Sequencers are needed to regulate flow, compensating for differences in the speed or volume of adjacent processes.

F14. The sequencer system and logic create the flow as they ask to be replenished as needed. Sequencer pulls a trigger for more iron, giving internal or external suppliers sufficient lead time to respond.

F15. Sequencer logic within area computer software controls the execution system. The execution system orders parts within a precise window of time to allow just-in-time delivery by internal or external suppliers.

F16. Manufacturing Resources Planning (MRP II) is reserved for long-term shop and supplier planning of rough material, human resources, and so on. It should not be used to schedule the shop. It supervises the sequencer-pulled quantities.

F17. There is very little or no buffer inventory.

F45. Iron in the continuous flow factory could be segregated in this manner:
- 40% controlled by suppliers and delivered to the assembly line or assembly sequencer

- 20% controlled by minimum inventory logic
- 30% controlled by the pull trigger
- 10% controlled by MRP

F46. An area computer serves a minifactory, creating and synchronizing flow using the pull-trigger concept. It uses the intelligent sequencer algorithm (Figure 4). The area computer concept allows data to stay at the lowest possible level.

F47. A continuous flow factory must have full-scope installation including computer-controlled material handling.

K. Integration

R9. Standardization is necessary to integrate now and prepare for future integration. Major areas of standardization include the MRP, CAD/CAM work stations, engineering graphics system (3-D), the MAP protocol, and corporate specifications for integration.

R10. A continuous flow factory information system links engineering, logistics, and the factory floor with a common data base called the "operation data base" which can be built up from existing data bases.

R11. Corporate specifications for factory integration must be developed in order to define the information flow between systems and prepare for future integration of data bases and systems. Vendors must follow the MAP protocol.

R12. A continuous flow factory uses modular software that is transportable between business units.

F41. There are three levels of computing in a continuous flow factory. Level 1 is at the process (the cell or FMS). Level 2 is the area computer, the centerpiece of the continuous flow factory. Level 3 is the plant host computer.

F42. Software for the continuous flow factory is based on the following principles:
- It is modular and transportable.
- It is under central control.
- Standardization has taken place.
- Systems serve a factory layout that has good flow.
- Scheduling is performed by the pull trigger at the part level. It is an electronic Kanban.
- Shared systems are avoided when possible.
- Data is kept at the lowest possible level.

F43. Basic software in the continuous flow factory includes material handling systems, cell or FMS controller systems, and area computing systems. There are vendors for modular software that performs these functions.

F44. Area computer software must work effectively in a multi-vendor environment. Integrating multiple systems is a specialized task which may require an outside integrator.

L. Human Resources

F26. Employee pride is generated through greater autonomy, broader-scope jobs, and personal accountability for quality achievement.

F27. Job content is broader, permitting more efficiency, competence, and coordination. It requires extensive training.

II. APPLYING THE RULES AND PRINCIPLES: ACTIONS AND GAINS

This section translates the rules and principles of continuous manufacturing into actions and estimates the potential gains to be realized. A tool for quantifying gains is presented in Chapter 9.

A. Strategy

Rules or principles	Actions
R1–3	Implement mandatory corporate cost reduction effort.
	Simplify administrative processes.
	Implement offshore sourcing strategy.
	Reorganize into profit centers, placing a high priority on performance.

Gains:

- Cost reduction programs lower total plant costs by 5% per year for 3 years.
- Process simplification reduces costs by 2% per year for 3 or more years.
- Offshore sourcing reduces total plant costs by 2%.
- Cost reduction associated with profit center organization is not quantifiable, but has a positive impact on profit.

B. Rationalization

Rules or principles	Actions
R4–6	Rationalize product line based on market survey.
	Consolidate excess space and capacity.
	Increase outsourcing, producing internally only what cannot be purchased from high-quality, cost competitive suppliers.

Gains:

- Impact of marketing analysis is not quantifiable, but has positive impact on sales.
- Impact of space and capacity consolidation is not quantifiable, but has positive impact on period costs.
- Impact of increased outsourcing is not quantifiable, but has positive impact on costs.

C. Purchasing and External Logistics

Rules or principles	Actions
R7 F22–25	Consolidate supplier network, developing stronger relationships with a smaller number of higher-volume suppliers.
	Institute new practices (standardized packaging, bar code labels, delivery to point of use, on-site replenishment) with suppliers and/or third party logistics providers.

Gains:

- Consolidated supplier network reduces material costs by 10%.
- New supplier practices eliminate many jobs:
    1. 80% of outside quality control certification specialists
    2. 100% of unpacking, counting, receiving personnel
    3. 80% of material control personnel
    4. 80% of storekeepers—purchased finished
    5. 90% of fork truck drivers who deliver material from receiving to point of use

D. Manufacturability

Rules or principles	Actions
R8	Prior to modernization, simplify product designs using simultaneous product and process engineering; assure ongoing use of simultaneous engineering.
	Work in partnership with suppliers to lower costs of purchased parts by designing for manufacturability.

Gains:

Manufacturability efforts reduce variable costs by 4%.

- Made parts: 1% reduction in variable labor, 1% reduction in direct part
- Purchased parts: 2% reduction in direct part

E. Layout, Automation, and Integration

The actions that must be taken to apply the rules and principles of layout, automation, and integration are presented together in order to illustrate a full-scope introduction of a continuous flow factory. The gains have also been consolidated.

Rules or principles	Actions
F5 F35 F38–40	Use continuous flow layout principles: • Design flow to move in one direction. • Lay out machines in minifactories, each responsible for producing a subassembly. • Abandon central receiving and deliver materials to point of use.

Rules or principles	Actions
	• Store rough materials in sequencer. • Decentralize heat treating and chemical processing when possible. • Locate utilities outside of the flow. • Reserve unallocated space in strategic locations for future use.
F1–2	Automate with a reasonable level of proven technology.
F29–31	Evaluate alternative technology (cell, cellular system, FMS, transfer line) using selection criteria and return on investment analysis.
F28	Deploy older technology in simple cells with computer controlled material handling.
F3 F6 F9–11	Deploy flexible automation (with resident programs and tooling) into minifactories, supported by computer-controlled material handling systems.
F36–37	Take advantage of process innovations (flexible carburizer, head changer technology) to overcome challenges of automation.
F13–17 F46	Schedule production and regulate flow with an intelligent sequencer; base sequencer logic on pull-trigger concept; reserve MRP II for long-term planning purposes.
F45	Segregate iron: 40% controlled by suppliers, 20% by minimum inventory logic, 30% by pull trigger, 10% by MRP II.
F11–12 F18–21	Integrate quality control into the process; manage quality with process capability and process control; stop flow for quality problems.

F32–34   Use flexible assembly systems (stall building, kitting, AGV feeding, flexible welding).

R9   Standardize MRP, CAD/CAM work stations, and three-
R11   dimensional software; issue corporate specifications for integration and use MAP protocol.

R10   Integrate engineering systems, MRP, and factory floor systems using a common operation data base.

F41   Implement three-level information system hierarchy (process computers, area computers, host computer).

F42–F43   Develop modular, transportable software for all levels of hierarchy; maintain centralized control, avoid shared systems, and keep data at the lowest possible level.

F44   Use outside integrator to develop area computer software that works effectively in a multi-vendor environment.

F47   Pursue full-scope installation strategy, with initial focus on integration of process and area computers.

F26–27   Broaden employee job content; give full accountability for quality and quantity to operators; implement comprehensive training program for all employees.

Gains:
1. Direct labor costs
2. Indirect labor costs
3. Indirect materials and expenses
4. Production materials costs
5. Inventory costs
6. Ease of management
7. Responsiveness to customers
8. Human resources

1. Direct labor costs

- Flexible automation eliminates set-up labor, reducing total direct labor costs by 20–30%.
- Resident tooling and programming reduces tool handling labor, lowering direct labor costs by 2%.
- Cellular systems, FMSs, and head changers reduce number of operators and direct labor costs by 50%.
- Stall building, kitting, and AGV feeding reduce assembly labor costs by 50%.
- Flexible welding systems reduce welding labor costs by 30%.

2. Indirect labor costs

Flow layout, flexible automation, flexible assembly, logistics, and integrated information systems permit the elimination of many jobs:

- Material control: 75%
- Production control: 75%
- Accounts payable: 70%
- Tool crib: 60%
- Work standards: 100%
- Quality control: 100%
- Scheduling: 100%
- Factory accounting: 100%

3. Indirect materials and expenses

Flexible automation reduces maintenance costs by 50%:

- Vehicles: 90%
- Machine tools: 50%
- Durable tooling and fixtures: 60%
- Building (compact layout): 25%

Integrated quality, process capability, process control and stopping flow for quality problems reduce cost of non-quality:

- Scrap: 70%
- Rework: 90%
- Warranty: 20%

Compact layout reduces utilities cost by 25%.

4. *Production materials costs*

- Computer aided plate nesting improves usage of steel plate from 80 to 95%, reducing rough material costs by 1%.
- Standardizing steel plate size and reducing the number of metallurgical specifications reduces rough material costs by 1%.

5. *Inventory costs (direct material and tooling)*

- Processing parts from start to finish and regulating flow by computer reduce in-process inventory from 90 days to 5 days.
- Migration of tooling cuts tooling inventory by 30%. (Tooling costs can be as high as 50% of machinery.)
- Flow layout, heat treat sequencer, pull-trigger concept, segregation of iron, and integrated information systems cut total inventory from 3.5 months of usage to less than 1 month, increasing the number of annual inventory turns from 3.4 to 12.

6. *Responsiveness to customers*

Flow layout, flexible automation, flexible assembly and welding, sequencer logic, area computing, segregation of iron, and integrated information systems allow:

- Reduction in throughput time (time of order to time of shipping) from 30–90 days to 5–10 days
- "Build-to-order" manufacturing, reducing prime product inventory from 30 days to 5 days
- More flexibility regarding product options. Impact on sales not quantifiable

- Fast changeover for new product manufacturing, reducing introduction cycle from 2½ years to 18 months

7. Ease of management

Computer controlled flow eliminates scheduling conflicts and crises, freeing supervisors to manage. Key benefits include:

- Less paperwork.
- No conflicting priorities for resource allocation.
- Natural production flow, easy-to-monitor production status.
- Lean, three-level organization cuts management costs by 40%.
- Readily available, highly accurate cost information.
- Reasonable automation and vendor support reduces start up time and cost.

8. Human resources

Broad job content, greater autonomy and accountability, and comprehensive training:

- Improve motivation and morale.
- Enhance employees' knowledge and skill levels.
- Develop a work force that is qualified to operate and manage a continuous flow factory to achieve maximum gains.

III. SUMMARY

Key points of the chapter:

- The rules and principles of continuous flow manufacturing can be recapped by grouping them in categories that include strategy, rationalization, purchasing and external logistics, design for manufacturability, layout, flexible automation, equipment selection, quality management, flexible assembly and welding systems,

internal logistics systems, integration, and human resources.
- All rules and principles can be translated into actions which produce quantifiable and nonquantifiable gains.

9

Consolidating the Gains of Continuous Flow Manufacturing

Chapter 8 outlined many gains that can be realized by applying the rules and principles of continuous flow manufacturing. This chapter presents a tool for consolidating those gains to determine the overall impact on total plant costs.

I. DEFINITIONS

A. "As Is" Versus "To Be"

To assess the impact of modernization on total plant costs, we must compare the cost of producing a product in traditional and continuous flow factories. We use the term "as is" to describe the cost situation in a traditional factory (Figure 35) prior to modernization. The "to be" situation represents the cost of producing the same volume of production in a continuous flow factory (Figure 3) with flexible automation, logistics, and systems.

B. Variable Versus Period Costs

Total plant costs are made up of two different types of costs: variable and period. Variable costs change (or vary) depending on the level of production. They increase or decrease in direct proportion to an increase or decrease in volume. Variable costs can be adjusted by management as business conditions and production volume change. Examples of variable costs include direct material, direct labor, assembly, energy, scrap, rework, and materials such as perishable tooling and weld rods.

Period costs are relatively fixed. They do not vary significantly with production volume and are more difficult for management to adjust if business conditions change. Examples of period costs include finance, purchasing, and information systems employees; depreciation; and the cost of heating or air conditioning. Separating total plant costs into variable and period costs permits budget monitoring regardless of changes in production volume. If a budget is established for given level of production (V), and volume changes (V^1), the difference $[(V^1 - V) \div V]$ is the volume change factor. This factor is a multiplier which allows variable cost to be flexed to the new volume. Period costs can then be added to permit an accurate comparison of total costs, excluding the effects of volume. In this formula, the volume of production is computed with a selected variable such as standard hours of production.

As we will see in this chapter, variable costs can be decreased dramatically in a continuous flow factory. Period costs can also be reduced, although to a lesser degree. As a result, a continuous flow factory has a higher proportion of period costs relative to variable costs than a traditional factory does. This means that although the continuous flow factory is more productive and cost effective than a traditional factory, it is also more sensitive to business cycles. It is also important to note that although significant cost reduction can be achieved through modernization, some costs will actually increase. Training costs may rise by about 20%. Labor costs for electricians and electronics personnel can increase by 30%. The cost of software maintenance may increase about 25%, and programming costs can double from their current level. Depreciation costs will also increase, depending on the level of investment. Despite these increases, the overall reduction that can be achieved through modernization makes continuous flow manufacturing a good investment and a sound business strategy.

II. PLANT COST STRUCTURE

To understand the impact of modernization on total plant costs, it is necessary to outline a typical plant cost structure. The numbers in the following table are based on manufacturing a sophisticated product and must be modified to reflect actual costs in the plant being modernized. The cost factors shown in the far right column will be used in the next section; the numbers in parentheses are for reference purposes only.

Total plant cost structure (as is)	Percent of total costs	Cost factor	
Variable costs			
Production material (purchased)	50	(1)	0.5
Labor (production)			
Operators	5	(2)	0.05
Others	3	(3)	0.03
Assembly	6	(4)	0.06

Total plant cost structure (as is)	Percent of total costs	Cost factor	
Logistics	5	(5)	0.05
Labor/material control handling			
Indirect material and expenses			
Freight			
Burden	7	(6)	0.07
Indirect material and expenses			
Energy			
First line supervision			
Scrap and rework			
Total variable costs	*76*	*0.76*	
Period costs			
General overhead	6	(7)	0.06
Supporting divisions			
Salaried/management payrolls			
Logistics	4	(8)	0.04
Assembly	2	(9)	0.02
Burden	12	(10)	0.12
General maintenance			
Indirect material and expenses			
Depreciation of assets			
Total period costs	*24*	*0.24*	

As the example indicates, three-quarters of total plant costs
are variable and one-quarter are period. It is also important
to note that direct labor costs represent just 5% of total costs,
so any effort to automate and modernize a plant must be
aimed at reducing more than just the cost of direct labor.
Modernization must lower a broad spectrum of variable
costs. It should also affect period costs. Achieving cost reduction in all areas of the business requires fundamental
changes in operating principles and a multidisciplined
approach to modernization.

III. ASSESSING THE GAINS

This section illustrates how each element of the cost structure shown in the previous chart is affected by modernization.

A. Variable Costs

1. *Production material*

As is	To be	Gain	Cost factor	Total plant cost gain
100	90	10	(1) 0.5	5%

Key reasons for reduction:

- Fewer parts made internally
- Smaller supplier network with higher volume suppliers
- Design for manufacturability
- Supplier manufacturability
- Standardization of steel plate sizes
- Computerized steel plate nesting

2. *Variable labor-operators and others*

As is	To be	Gain	Cost factor	Total plant cost gain
100	50	50	0.08 (2) + (3)	4%

Key reasons for reduction:

- Flexible automation with no set-up time and resident tooling
- Continuous flow layout concepts

- Selection of the right level and type of automation for job
- Flexible welding systems
- Integrated automation, material handling and logistics

Note: 50% reduction is an aggressive example. The average is 30 to 40%.

3. *Variable assembly*

As is	To be	Gain	Cost factor	Total plant cost gain
100	60	40	(4) 0.06	2.4%

Key reason for reduction:

- Flexible assembly concepts (kitting, AGV feeding, computer controlled logistics)

4. *Logistics*

As is	To be	Gain	Cost factor	Total plant cost gain
100	80	20	(5) 0.05	1%

Key reasons for reduction:

- Efficient external logistics practices (point-of-use delivery, bar coding, standardized packaging, third party logistics providers)
- Internal logistics systems based on pull trigger concept
- Integrated automation, material handling, and logistics

5. *Variable burden*

Cost	As is	To be	Gain	Cost factor	Total plant cost gain
Energy	100	90	10		
First line supervision	100	50	50		
Indirect material	100	80	20		
Scrap/rework	100	50	50		
Total	100	70	30	(6) 0.07	2.1%

Key reasons for reduction:
- Ease of management
- Quality focus

B. Period Costs

1. *Period labor*

As is	To be	Gain	Cost factor	Total plant cost gain
100	50	50	0.08*	4%

* Includes portions of cost factors (7), (8) and (9)

Key reasons for reduction:
- Computer controlled material handling
- Assembly material control
- Welding material control
- Pull trigger concept
- Area computing
- Integrated systems
- Reduction of quality control labor
- 70% or more reduction of production control labor (work standards, scheduling, factory accounting)

2. *Utilities*

As is	To be	Gain	Cost factor	Total plant cost gain
100	75	25	0.05*	1.25%

* Includes portion of cost factor (10)

Key reasons for reduction:
- Consolidated manufacturing space
- Compact layout

3. *General maintenance*

As is	To be	Gain	Cost factor	Total plant cost gain
100	75	25	0.06*	1.5%

* Includes portion of cost factor (10)

Key reasons for reduction:
- Flexible automation and new equipment, reduced amount of building maintenance needed (fewer buildings, compact layout), and fewer maintenance vehicles required

4. *Repair maintenance*

As is	To be	Gain	Cost factor	Total plant cost gain
100	50	50	0.03*	1.5%

* Includes portion of cost factor (10)

Key reasons for reduction:
- Flexible automation and newer equipment improve machine reliability and uptime, reduce maintenance machine tools, and reduce maintenance in tool room.

5. *Depreciation*

As is	To be	Gain (loss)	Cost factor	Total plant cost gain (loss)
100	120	(20)	0.05*	(1%)

* Includes portion of cost factor (10)

Key reason for cost gain:

- Major investment in new assets

IV. TOTALING THE GAINS

When the gains for each element of the cost structure are totaled, we determine that implementing a continuous flow manufacturing strategy can reduce variable costs by 14.5% and period costs by 7.25%.

Variable costs

Production materials	5%
Labor	4%
Assembly	2.4%
Logistics	1%
Burden	2.1%
Total variable	14.5%

Period costs

Labor	4%
Utilities	1.25%
General maintenance	1.50%
Repair maintenance	1.50%
Depreciation	(1%)
Total period	7.25%

Total plant cost reduction   *21.75%*

This figure (21.75%) does not include nonquantifiable gains discussed in Chapter 8, such as reorganization into profit centers, rationalizing the product line, and rationalizing manufacturing capacity. By including these nonquantifiable benefits, a total plant cost reduction target of 22% is realistic and achievable

V. LEVERAGING THE GAINS

As discussed in Chapter 8, the benefits of a modernization strategy can be leveraged by combining it with defensive strategies.

Defensive strategies (squeezing existing cost structure)

Cost reduction program (5% per year for 3 years)	15%
Process simplification (2% per year)	2%
Offshore production (4% in direct material costs × 0.5)	2%
Total defensive strategies	19%
Continuous flow strategy (changing cost structure)	22%
Total possible objective	*41%*

VI. OTHER BENEFITS

In addition to the permanent changes in cost structure that can be achieved through modernization, many other benefits can be realized. As noted in Chapter 8, other benefits of continuous flow manufacturing include:

- Allowing capacity adjustment and plant closing.
- Reducing direct material inventory from 3.5 months of usage to 1 month.
- Increasing inventory turns from 3.4 per year to 12. This one-time-only gain improves cash flow and return on investment.

- Reducing tooling inventory by 30%.
- Reducing direct labor and indirect period labor by as much as 50% permits a lean organization design.
- Cutting total lead time (factory plus suppliers) from 150 days to 45 days.
- Reducing throughput time from an average of 60 days to 6 days.
- Improving responsiveness to customers (build-to-order flexibility, shorter new production introduction cycles, capacity to grow rapidly, strong positive image of modern manufacturer). These advantages should translate into incremental sales and improve revenues.

VII. SUMMARY

Key points of the chapter:

- To assess the impact of modernization on total plant costs, the "as is" cost scenario must be compared to the "to be" situation for the same volume.
- Variable costs vary in direct proportion to production levels, while period costs are relatively fixed. Costs are split into variable and period to improve budgeting and monitoring accuracy.
- Direct labor costs represent only 5% of total plant costs. Therefore, to generate an acceptable return on investment, a modernization project must reduce a broad range of variable and period costs.
- Continuous flow manufacturing can cause a major reduction in variable costs (14.5% in the example in this chapter) and a significant reduction in period costs (7.25% in this example).
- The gains of continuous flow manufacturing can be leveraged by applying defensive strategies such as cost reduction, process simplification, and offshore sourcing.

- Some costs (training, electricians, programming, and depreciation) will increase when a continuous flow manufacturing strategy is implemented.
- Although continuous flow factories are more sensitive to changes in the business cycle than traditional factories, the total benefits of modernization make it a wise investment and a good business strategy.

Justifying, Approving, and Monitoring the Investment

This chapter discusses the financial justification process for a continuous flow factory. It also covers how modernization proposals are approved and the major factors that must be monitored as the investment is implemented.

I. CRITICAL COMPONENTS OF THE FINANCIAL JUSTIFICATION PROCESS

Following are the critical components that must be addressed in the financial justification of a continuous flow factory.

A. Vision

The justification must be based on a clear vision of the proposed investment. The vision must be well communicated throughout the organization and strongly supported by those who will implement it. Developing and communicating a clear vision of the modernization program can be challenging because the concepts are fairly complex and there are few factories to visit. People who do not understand the total vision tend to focus on one aspect of it, such as just-in-time delivery or flexible automation. If the vision is not fully understood and implemented in its entirety—with a new layout, flexible automation, and integrated logistics and systems—the full benefits of continuous flow will not be realized.

B. Minifactories

It is easier to plan, justify, implement, and manage a major modernization program when a total plant is divided into small entities or minifactories. Management can divide the plant in any way, but the concept works best when every entity produces a well-defined subassembly, an identifiable piece of engineering with a unit cost and a market price. Establishing individual, self-contained minifactories within a block layout permits more accurate budgeting. It also speeds the implementation process and allows more precise cost monitoring.

C. "As Is/To Be" Approach

Modernization programs should be justified using the "as is/to be" approach in which current ("as is") costs and inventory levels are compared to future ("to be") costs and inventory. Accountants can determine the "as is" cost structure with a high level of precision once the minifactories have been well defined. It is more challenging to determine the "to be" situation. "To be" numbers reflect the implementation of a total vision and must be arrived at with input from the entire organization. There must be a consensus on the numbers and a strong commitment to meeting or surpassing them. Educating the organization at every level will help build support for the vision and projected goals. In addition, progressive reviews of the individual projects that constitute the total modernization program—first during the planning phase at the plant management level, then during the approval phase at the corporate level—will help generate understanding of and commitment to the goals.

D. Net Capital Outlay

When determining the amount of capital needed for modernization, two factors must be deducted from total capital: replacement capital and proceeds from the sale of surplus or obsolete equipment.

Replacement capital refers to the money that would have been spent to replace worn-out equipment and obsolete machines even if there were no modernization project. When replacement capital is subtracted from total capital, the remaining number reflects the incremental capital needed for modernization. The second deduction that must be made from total capital is the value of proceeds from the sale of surplus and obsolete equipment. The final figure (net capital outlay = total capital – replacement capital – proceeds from sale of surplus equipment) is the most accurate number to use in return on investment calculations.

Two things to remember about replacement capital and proceeds from surplus equipment:

- The higher the replacement capital, the easier it is to justify a project. To prevent plants from overstating future replacement capital needs, it is best to have an independent corporate staff person approve replacement capital numbers using historical data.
- When surplus equipment is sold it normally brings more than book value, especially when accelerated depreciation is used. This could lead to a capital gains tax liability unless careful attention is paid to the timing of the sale and the new acquisition.

E. Start Up Costs

Start up costs will be incurred as the plant is rearranged into minifactories and new equipment and systems are ramped up. Identifying start up costs improves budgeting accuracy and allows more precise measurement of results.

Key components of start up costs are:

- Rearrangement expenses
- Amount of time the old and new equipment are running simultaneously
- Time required to bring new equipment to expected level of production
- Time required to reduce employment to planned levels
- Training costs

Experience indicates that start up costs are often underestimated. It is important to budget them carefully as they affect the overall return on investment. Start up costs can be minimized by eliminating unnecessary complexity.

F. Inventory

Future inventory levels play a key role in financial justification, affecting future cash flow. A continuous flow factory can operate with dramatically reduced inventory levels. Although the reductions are aggressive, they are realistic and achievable as long as the total modernization plan includes a flow layout, flexible automation, computerized material handling, and integrated systems.

G. Timing

The length of time required to complete implementation and begin realizing the cost and inventory benefits has a major effect on investment justification. Discounted cash flow analysis and return on investment calculations are extremely sensitive to timing, so much attention should be paid to this issue.

Experience shows that people can accurately forecast the new cost structure and surpass their inventory goals, but they generally underestimate the time required for implementation. Implementation time is often extended because the projects are unnecessarily complex. To speed implementation and enhance ROI, every effort must be made to simplify projects. In addition, a sensitivity analysis should be completed so management understands the impact on ROI if the project is extended by six months or one year.

II. CORPORATE-LEVEL JUSTIFICATION

Justification for a major modernization program begins at the top of the organization. This is essentially a feasibility study, assessing:

- The impact of the investment on corporate operational results
- The effect on corporate cash flow
- The expected return on investment (ROI) and long-term return on assets (ROA)

These issues and others can be answered by developing a macro financial model which consolidates all "as is" and "to be" numbers, based on input from plant feasibility studies.

A. Impact on Corporate Operational Results

Figure 42 shows a simple bar chart that projects the incremental before-tax profit due to modernization. It considers all factors: cost and inventory reductions at the time they happen, incremental sales, the negative impact of higher depreciation, and start up costs during the early years. Clearly, the short-term effect on profitability is negative, but the long-term gains are positive.

B. Effect on Corporate Cash Flow

Figure 43 shows a 10-year picture of the effect of modernization on cash flow before taxes. This is a cumulative cash flow

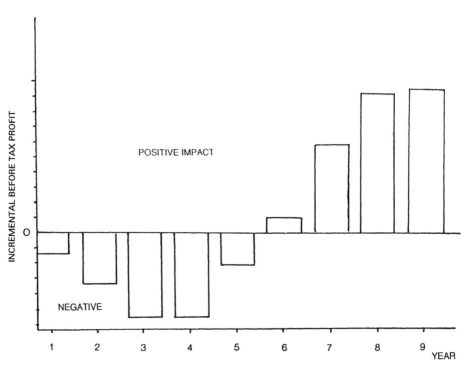

Figure 42 Incremental before-tax profit due to modernization.

chart; each year's cash flow is added to that of all previous years. This is a good tool for assessing the investment because cash flow includes both the resources needed for modernization and the gains to be realized from it. Resources needed for modernization (capital expenditures) are entered each year as negative cash flow. Start up costs are also included in negative cash flow. Positive cash flow includes inventory improvements; cost reductions; and increases in revenue due to faster throughput time, increased flexibility, and improved responsiveness to customers.

This analysis uses discounted cash flow techniques. All future benefits are discounted to present value, using a standard present value interest table. For example, if the cost of capital is 16%, benefits are discounted as follows:

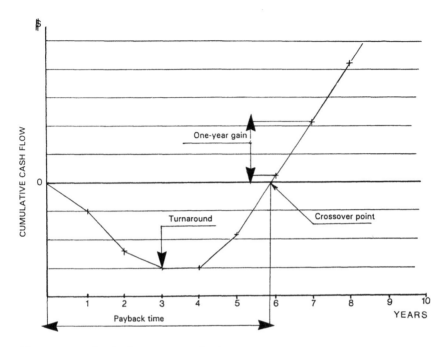

Figure 43 Cash flow—cumulative.

Year of gain	Discount factor
1	.86
2	.74
3	.64
4	.55
5	.48
6	.41
7	.35
8	.31

Obviously, the earlier the gain can be realized, the greater the positive impact on cash flow.

As the chart indicates, cumulative cash flow for modernization is negative and trends in the wrong direction for three to four years. Cash flow stabilizes at a turnaround point in about the fourth year. It reaches the zero line or crossover point after year six. From then on, using the desirable discount rate, the cumulative cash flow generated by the investment exceeds total cash outlay. This example, with a six-year payback, is conservative. Five years is achievable if the modernization program includes reasonable automation and systems.

There are many publications available on the discounted cash flow methodology. Two excellent ones are "Cost Justification," edited by the National Machine Tool Builders of America (1990), based on a study prepared by Ernst & Young, and "Investment Justification," published by Giddings & Lewis, Inc. (1989).

C. Performance Indicators (ROI and ROA)

To justify the modernization investment at the corporate level, top management normally requires an ROI calculation. ROI measures the level of benefits that will be realized by spending a given amount of capital. It is calculated as follows:

$$\text{ROI} = \frac{\substack{\text{Modernization gains} \\ \text{(cost and inventory reductions)}}}{\substack{\text{Net capital outlay} \\ \text{(total capital} - \text{replacement capital} - \\ \text{proceeds from sale of surplus equipment)}}} \times 100$$

The higher the ROI, the better the plan. A good ROI for a modernization proposal is between 30 and 35%, with a 5% contingency. Projects with an ROI of less than 25% should not be approved. As mentioned earlier, ROI is affected by timing. Delays related to complexity will increase start up costs and reduce the overall return.

Although ROI is a good tool for choosing an investment alternative, it will not measure how effectively the factory will run when all the new assets are in place. This requires a return on assets (ROA) analysis. ROA indicates how much profit can be generated with the money that is invested in the business. A good ROA is about 15%. It is calculated as follows:

$$\text{ROA} = \frac{\text{Yearly profit}}{\substack{\text{Net assets} \\ \text{(existing equipment + incremental} \\ \text{capital + all inventories)}}} \times 100$$

Modernization increases net assets and affects ROA. The increase in net depreciation that accompanies modernization has a significant effect on yearly profit, which also affects ROA. When management is evaluating the ROA of modernization, good business sense prevails. To justify modernization on an ROA basis, the long-term profit improvement that will be generated from continuous flow manufacturing must be high relative to all the new assets the company or business unit will carry.

III. PLANT-LEVEL JUSTIFICATION

The corporate financial model is a consolidation of all plant financial models. Each plant model is based on the "as is" and "to be" numbers from of its own minifactories. While the numbers in the corporate model are projections, those in the plant model should be firmer. The organization must be committed to achieving them. The plant financial model is a gate which must be passed through before capital expenditures can be approved for creation of a new minifactory.

The plant model measures the impact of modernization on plant profit and plant cash flow. It also projects ROI and ROA at the plant level. The plant financial model should also include market-based cost targets for each minifactory subassembly. A good ROI and ROA are not enough to justify modernization. It is equally important to achieve cost reduction that will make subassemblies competitive with market prices.

IV. APPROVING THE INVESTMENT

While corporate and plant financial models are useful tools for justifying a modernization investment, the actual approval of projects should take place at the minifactory level. Proposals submitted for approval must include all aspects of the minifactory—layout, automation, material handling, logistics, and systems. In other words, there should be no proposals for single machine tools or computer systems. Instead, each proposal should address the total costs and benefits of modernizing a minifactory. A three-step process for reviewing proposals has proven to be effective.

Step 1 is a concept review. This should be a relatively informal working session which takes place early in the process and involves the approver (the senior executive responsible for modernization) and the doers (a manufacturing engineer and team). In this session, the proposal is discussed openly. Ideas are shared, issues are raised, and concerns are expressed. Taking time to review and talk about the concept prior to developing the final proposal improves the quality of the final proposal. It also prevents the cost and disruption that occur when final proposals are developed, but rejected or returned to the organization for additional work.

The type of review held at Step 2 depends on the outcome of the concept review. If it produced many differences, an interim review between the approver and the doers may be required, followed by a final review with the approver. If the concept review went smoothly, a final review with the approver is all that is necessary.

Step 3 is the corporate review. At this stage, a mature proposal is presented to those who must approve the capital—the top managers of the company and the board of directors.

V. CHECKLISTS FOR ASSESSING A PROPOSAL

The following lists can be used by management during the approval process to assess the many critical aspects of a modernization proposal. The lists can also be used during the development phase of a proposal. They are generic enough to leave room for innovation and variation of the vision. Publishing the lists and disseminating them broadly helps communicate the vision and principles of continuous flow manufacturing. It also improves the quality of final proposals and should speed the approval process. These lists are intended as broad guidelines only and can be modified to reflect a company's unique experiences.

A. Prelude to Modernization

This review should take place once in each plant that will be modernized.

- Has a product strategy been finalized?
- Have target costs been identified for each product?
- Is capacity sized appropriately and conservatively?
- Is there a plan for reducing excess manufacturing space?
- Has a comprehensive make/buy review taken place?
- Are we committed to buy as many parts as possible from the world's lowest cost, highest quality suppliers?
- Do we have evidence that the parts or subassemblies we will make cannot be purchased?
- Has the number of suppliers been reduced significantly?
- Has product design been simplified?
- Is design simplification ongoing?
- Are we using simultaneous product and process engineering?
- Are products being designed for manufacturability?
- Have we reduced the number of part numbers per product?
- Have we consolidated steel codes?
- Are we using fewer material specifications?
- Are engineering standards for dimensions and tooling being used?

B. External Logistics

This review should take place once in each plant that will be modernized.

- Has a packaging standard been issued to suppliers?
- Have suppliers accepted packaging and bar code standards and do they have a target date for 100% compliance?
- Is just-in-time delivery being implemented?
- Are parts and materials being delivered to multiple points of use?
- Has a system been finalized for implementing the pull trigger concept with suppliers?
- Do suppliers understand the respective roles of the MRP II (long-term planning) and the pull trigger (execution)?
- Has supplier training taken place?
- Is a supplier certification program in process?
- Is inspection taking place at the supplier's business?
- Has traffic for lines of supply been finalized?
- Have we analyzed the economics of using a third party logistics provider?
- Has auditing accepted the third party logistics concept?
- If third party logistics providers are used, have we developed procedures for quantity checks and matching receipts with purchase orders for payment?
- Is a logistics provider delivering to assembly locations?
- Does the logistics provider have a pull trigger system that notifies suppliers to replenish its warehouse?
- Does the payment procedure for suppliers who deliver to the logistics company satisfy auditing?

C. Layout

These questions should be answered in a single review of each plant that will be modernized.

- Which direction is flow moving?
- Has a block layout with minifactories been developed?
- How many minifactories have been created?
- Where are the minifactories located?
- Where are utilities located?
- Where are lockers, restrooms, and cafeterias located?
- Where is the factory office located?
- Is there sufficient unallocated space for growth?
- Where is the unallocated space located?
- Where are the sequencers located?
- How large are the sequencers?
- How was sequencer size calculated?
- How are the processes fed?
- How is the assembly line fed?
- What is the path for future expansion?
- Which processes (heat treat, chemical processing) take place in a centralized location?
- If centralized processing is being done, what is being done to protect the flow?
- What type of material handling systems are being used?
- What alternative material handling systems have been considered?
- Are main process and assembly stores located in the right place?
- Has material handling been simulated and minimized?
- Is the layout free of all islands that interrupt flow?

D. Minifactory

This technical and detailed review should be completed once for each minifactory.

1. General

- Is the minifactory a well balanced mix of logistics, material handling, automation, and systems? Is there

too much emphasis on machinery over logistics and systems?
- What vendors have been consulted? Have any been overlooked? Are more trips to suppliers necessary?
- How will iron be machined (cells, cellular systems, FMS, flexible transfer)?
- Why has the iron been segregated in this manner?

2. Layout

- How does the iron flow?
- How are the processes fed?
- How is assembly fed?
- What material handling devices are being used?
- Where are sequencers located?
- Have sequencers been sized appropriately?
- What system activates the flow?
- Are machines, sequencers, and assembly located to minimize material handling?

3. Primitive cells

- Has an honest effort been made to reuse existing equipment?
- Have employees participated in the layout of their work stations?
- Is each cell a full-scope operation including washing, deburring, and subassembly?
- How many operators will run the cell?
- What is the process for quality management?
- What is the process for perishable tool setting?
- How is the cell fed?
- How is production pulled from the cell to assembly?

4. Cellular systems

- What machinery has been selected and why?
- How are programs downloaded from the area computer?
- What material handling devices are being used?

- How many operators are needed?
- What is the utilization rate of the system?
- Why has a cellular system been chosen over an FMS?
- What is the difference in ROI between a cellular system and FMS?
- Is floor space a problem?
- Has the cell been integrated with the sequencer, material handling systems, and area computer?
- How is quality being managed?

5. FMS

- Which supplier has been selected and why?
- Which other suppliers were considered?
- Why has an FMS been chosen over a cellular system?
- How many part numbers will be run?
- Has iron been segregated into families to run on the FMS?
- Can we run a simple FMS for large families of parts?
- Can the FMS operate unattended? For how long?
- Has the FMS been programmed to stop if there is a problem?
- How many pallets are required and at what cost?
- Have we selected the simplest technology that will do the job?
- Are we implementing tool management and delivery systems?
- Will we use tool compensation?
- Are we capable of a full-scope introduction?
- Has the FMS been integrated with the sequencers, material handling, and area computer?
- What is the process for quality management?

6. Flexible transfer lines

- Why has flexible transfer been chosen over FMS?
- How stable is our part design?
- What is the required investment in heads and pallets?

- How many parts will be run? (Do not accept more than six.)
- Do we have the capability to introduce new parts?
- What is the capacity of the head magazine?
- Has the flexible transfer line been integrated with the sequencers, material handling, and area computer?
- Is unattended operation possible? For how long?
- What is the process for quality management?

7. Dedicated transfer lines

- Why has dedicated transfer been chosen over flexible transfer?
- Is there evidence of capital savings over flexible transfer?
- Is there evidence of long, stable part life with no major design changes?
- How many parts will be run?
- Has the dedicated transfer line been integrated with the sequencers, material handling, and the area computer?
- Is unattended operation possible? For how long?
- What is the process for quality management?

8. Equipment

- Is equipment robust, state of the art, and low risk with no unnecessary gadgets?
- Is it flexible to engineering changes?
- Does it incorporate simple, user-friendly systems?
- Can it be easily integrated into factory area?
- Does good documentation exist to support equipment, maintenance, and software?
- Are spare parts readily available?

9. Vendors

- Which vendors have been consulted?
- Should others be considered?

- Should additional trips to suppliers be planned?
- Are vendors well known and financially strong?
- Is staff available to assist with vision, justification, budget, and training?
- Do we have access to top management?
- Do vendors have reputation for pricing surprises?
- Can delivery timetables be met?
- Can vendors help with foundation, cooling system, and perishable and durable tooling programming?
- Do vendors have a history of responsiveness to customers?

10. Machine Design

- Do machines feature rapid traverse rates (2000 inches per minute)?
- What is the spindle speed? How does it compare to competition?
- Is spindle speed fast enough to take new coated tools?
- Does machine use digitally pulsed (computerized) drives for spindle speed control?
- Does machine have thermal stability?
- Are tools rigidly held?
- Is tool durability known?
- Do electrical components have good contamination control?
- What is the reliability of electrical components?
- Is there a plan for preventive maintenance that includes a list of replacement components?
- Has the use of electrical components been minimized and computer controls maximized?

E. Systems

1. Plant level

- What is the status of standardization of CAD/CAM, 3-D design, and work stations?

- Does plan comply with corporate selection of MRP II? Are there problems adapting to corporate standards?
- Which corporate modular software will be transported? What local enhancements are necessary? At what cost?
- Has the material handling, cell, and area computer basic software been reviewed in detail?
- Who will integrate the area computers with the host? What is the cost of the hardware, software, and network?
- What is the status of the operation data base build up?
- How many area computers will be used? How many minifactories?
- Has the hardware hierarchy been reviewed?
- Have the roles of IBM and DEC been defined?
- Has the iron been segregated? What percentage is scheduled by MRP? By minimum inventory? By the pull trigger? Delivered to assembly by suppliers?

2. *Minifactory level*

- Has the cost of area computers, software, and network been included in the project?
- Who will integrate at the area level (machinery, material handling, sequencer)? What funding is required?
- Do all vendors comply with the MAP and corporate specifications for integration?
- Does the information flow result in continuous, high-quality production?

VI. MONITORING THE INVESTMENT

Once the investment has been approved and is being implemented, monitoring becomes critical. Monitoring is necessary to determine:

- What is the current status of the execution of the plan? How is the installation of new equipment and systems progressing?
- What results are being achieved? What progress are we making toward reaching the financial goals of the project?

These issues are closely linked because timing has a direct effect on results. Delays in implementation reduce the return on investment. The organization and equipment vendors must understand the link between timing and results to assure that proposals are developed with realistic timetables and implementation proceeds at the fastest rate possible.

A. Monitoring Execution of the Plan

Each plant should provide a regular report on the status of execution. Factors to report include:

- Amount of plant rearrangement completed (percent of total)
- Amount of capital spent (percent of total projection)
- Assets in place and working in new location (percent of total)
- Number of suppliers integrated (percent of total)

B. Monitoring Results

Plants should also provide regular reports of results being achieved. Numbers should be collected and reported by minifactory, then consolidated at the plant level. Key factors for the report are:

- Employment reduction (adjusted for volume)
- Reduction of in-process inventory
- Product cost reduction
- Reduction of total plant costs
- Reduction of throughput time in days
- ROI by minifactory

C. Monitoring Corporate Results

Plant results can be consolidated quarterly to provide an overall assessment of results being achieved. Key factors in the report include:

- Total employment reduction (adjusted for volume)
- Total reduction in in-process inventory
- Total reduction of product costs
- Average reduction in throughput time
- Overall return on investment
- Impact on profitability (Figure 42)
- Impact on cumulative cash flow (Figure 43)

VII. SUMMARY

Major points of the chapter:

- Justifying a continuous flow factory requires a comprehensive analysis that embraces all aspects of the vision. The accounting tools for justification exist. The challenge is to assure the quality and scope of the input data.
- The "as is/to be" methodology is the most appropriate way to assess the total financial impact of a modernization investment.
- Calculations for return on investment should be based on net capital outlay. Net capital outlay = total capital – replacement capital – proceeds from the sale of surplus equipment.
- Replacement capital is the amount of capital that would have to be spent to replace worn out or obsolete equipment if there were no modernization program.
- Start up costs penalize the project, slowing down improvements and reducing the return on investment. Start up costs can be minimized by using proven technology and reducing the complexity of the project.

- To avoid the cost of running old and new systems simultaneously, old equipment should be declared surplus and sold.
- Financial modeling at the plant and corporate levels allows a critical assessment of the modernization project.
- A corporate feasibility study assesses:
  1. The impact of modernization on corporate profitability.
  2. The impact on corporate cash flow. (Discounted cash flow techniques should be used. Cumulative cash flow should be measured to project the turnaround point when cash flow becomes positive and the crossover point when total cash inflow exceeds total cash outflow.)
  3. The overall return on investment to the corporation (minimum acceptable ROI is 30 to 35% with a possible slippage of 5%).
- After modernization, the return on assets measurement should be used to determine how much profit is being generated by the assets employed in the business.
- The approval process includes:
  1. A concept review to share the vision and agree upon the direction of the project before committing too much time and resources.
  2. An interim review if there were many problems identified in the concept review, followed by a final review with the senior executive accountable for modernization.
  3. A corporate review presented as a request for capital to senior management and the board of directors.
- Checklists have been developed to help assess a proposal. These lists, which are also useful for communi-

cating the principles of modernization and developing a quality proposal, should be widely distributed.

- Systems must be developed to monitor how the plan is being executed and what results are being achieved.
- Monitoring should be done at the minifactory level, then consolidated at the plant and corporate levels.

11

Creating a Continuous Flow Factory

The challenges of creating a continuous flow factory in a brand new plant are primarily technical. Making new technology work productively and profitably is difficult and time-consuming. But it is a manageable job, and it is typically much easier than modernizing an existing plant into a continuous flow factory. When a traditional factory is modernized, there are many challenges in addition to those related to technology. People resist change, plant rearrangement is costly and disruptive, and it is difficult to maintain production levels when new machines and systems are being installed. This chapter discusses the challenges of converting a traditional factory to a continuous flow operation. It presents an overall conversion process and outlines a "how to" approach for some of the most important steps in the process. It also discusses mistakes to be avoided and keys to managing change.

I. THE CONVERSION PROCESS

Following is a process for converting a traditional factory to a continuous flow operation. The steps in this process are rarely completed in perfect order. Time pressures, budget constraints, resource limitations, and other circumstances dictate that some steps be completed simultaneously and others be eliminated entirely. However, for best results, all steps should be completed and the order of the process should approximate the one described.

A. Consensus on a Vision

Reaching agreement on a vision for the organization can take a year or more and involves several steps:

1. *Assemble a data-gathering team.* Appoint a team of three or four executives with engineering, manufacturing, and systems backgrounds. The group will visit leading manufacturers to gather data for the vision. The group must include innovators, risk takers, and champions of change.

2. *Visit the world's top manufacturers.* Tour as many leading-edge companies as is financially possible, including European and Japanese operations. Do not limit your visits to companies in your business. Competitors often use similar manufacturing methods, so new ideas are more likely to be found in organizations outside your industry. For example, the sequencing logic explained in Chapter 3 was learned from a chocolate candy company. The use of shuttle technology and large assembly FMSs

(described in Chapter 4) was discovered on a visit to a German builder of prefabricated homes. The purpose of these visits is twofold: (1) to identify principles which make the manufacturing operations successful, and (2) to determine how the principles can be applied and improved in your operation. It will be rare for you to copy another manufacturer's operation exactly, so it is important to focus on principles rather than techniques. Key areas to research include the general plant layout, the systems that command the flow, and the factory's basic operating principles.

3. *Appoint an architect.* A high-level executive should be appointed as full-time architect of the modernization program. This executive will have accountability for developing and executing the vision. He or she will review all capital projects and present them for approval to management and the board of directors. Corporate systems people should report to the architect.

4. *Conduct an off-site seminar.* To begin reaching consensus on a vision of modernization, an off-site seminar with key management people can be effective. Those who visited leading world manufacturing operations should attend, as well as plant managers, systems people, accountants, and representatives from the marketing organization. The goal of this meeting is to educate participants about continuous flow manufacturing, build a business case for modernization, and gain widespread support for pursuing this strategy.

5. *Create overall corporate vision and refine at the plant level.* Although the fundamental elements of the vision will result from the off-site seminar, it may take a year or more to develop a detailed corporate strategy and specific plans for each plant. Planning will take place at the corporate level and

within the plants. By involving in the planning stages those who will ultimately implement the vision, there should be less resistance to change and faster, more effective implementation.

B. Prelude to Modernization

As a detailed vision is being developed for each plant, work can begin on the prelude to modernization (Chapter 2). Taking these actions early in the process reduces the total time required to complete the modernization project.

1. *Set the product strategy.* Conduct market research to determine future product line. When product strategy is finalized, excess capacity can be identified, and plans can be made to size capacity at an appropriate level.

2. *Determine "core make" and "core buy" parts.* Appoint multi-functional teams of engineering, manufacturing, and purchasing employees to determine which parts and components will be made internally and which will be purchased from suppliers. The objective is to buy as much as possible to cut costs, save capital, and focus the new manufacturing operation on critical components that cannot be purchased cost effectively.

3. *Consolidate supplier organization.* When "core buy" components have been identified, make arrangements to purchase them in large volumes from low-cost, high-quality suppliers. The goal is to reduce the size of the supplier network and build stronger relationships with fewer vendors.

4. *Introduce design for manufacturability concepts and systems.* Simultaneous engineering principles and practices must be introduced to the organization and the supplier network. A plan should be developed to assure that in the long term, all "core make" and "core buy" parts will be designed for low-cost, high-quality manufacture.

5. *Begin disposing of assets to make room for rear-rangement.* As "core buy" components are identified, the assets that have been used to make them can be declared surplus. These assets, plus others which were declared surplus when the future product strategy was developed, may be sold or reused in the new continuous flow layout. Disposing of surplus assets will reduce costs, improve utilization, and enhance return on assets. It will also create sufficient space for a phased rearrangement of the facility into a continuous flow layout.

6. *Begin designing the continuous flow layout.* Key factors to consider during the preliminary development of the layout are direction of flow, space allocation and location of each minifactory, location of assembly area, sequencer sizes and locations, and the path for future expansion. Care should be taken to prevent islands or interruptions to flow. Simulation techniques should be used to minimize material handling.

7. *Set up a "war room."* A central room in which all meetings regarding modernization are held can be an effective way to create and disseminate the modernization vision. The walls of the war room can be used to illustrate principles and display flow charts, factory layouts, and other concepts. A scale model of the future plant could also be on display in the war room and used to inform employees and their families. A war room should be established at each plant being modernized, and at the corporate headquarters where it is used for identifying fundamental principles and the corporate systems strategy. One of the best war rooms in the United States was located at GE's Erie, New York, locomotive plant during the 1980s.

8. *Begin segregating iron into broad families for scheduling.* Teams of manufacturing and purchas-

ing employees can begin developing the logistics plan by dividing parts into categories based on how they will be scheduled. An example logistics plan segregates the iron in the following manner:

- 30% scheduled by the pull trigger. This includes "core make" parts which are produced when the pull trigger signals the process to complete a load of parts and deliver it to the sequencer. It also includes "core buy" parts which are produced when the pull trigger signals the supplier to make a load of parts and deliver it to the sequencer.
- 20% scheduled by minimum inventory logic. This is primarily low cost hardware which is produced and delivered to the sequencer when inventory reaches a specified minimum level.
- 40% scheduled, managed, and delivered to the point of use by suppliers. This is a sophisticated service in which suppliers monitor bins and boxes at the assembly line and stock them as needed, similar to the way vending machines are checked and replenished on an ongoing basis.
- 10% scheduled by MRP II and delivered as needed to the assembly area. Typical parts controlled by MRP II would be vehicle seats and batteries.

C. Detail Processing

Detail processing at the plant level requires teamwork. Engineering contributes by adapting product designs to the manufacturing process. Logistics people help determine how to feed the process or assembly sequencer, how to get just-in-time delivery to points of use, and what types of material handling devices are best. Systems people are also instrumental in creating the information systems that will move material continuously and synchronously. Key steps in detail processing:

1. *Design an external logistics plan.* This long process involves working with suppliers to determine packaging standards, bar code standards, just-in-time strategies, supplier certification requirements, lead times, traffic strategies, and the possible use of third-party logistics companies.

2. *Develop processing and internal logistics plans in each minifactory.* A small group of processors should work with engineering to define processes and design products simultaneously. If the product already exists, the challenge is to adapt it to the process without major re-engineering. The manufacturing process may be a primitive cell, cellular system, simple FMS, higher technology FMS, grouped FMSs, flexible transfer line, or dedicated transfer line. Material handling to feed and exit the process must also be identified. Sequencers should be linked to processes by computer controlled material handling. The criteria for selecting a process were discussed in Chapter 4. In general, primitive cells are good for reusing existing equipment, but have low utilization and use more space. Cellular systems and simple FMSs are highly recommended. Higher technology FMSs are good if a major cost reduction is needed. Dedicated transfer lines are only effective with a small number of very stable parts. Flexible transfer lines are preferred over dedicated because they are less apt to become obsolete quickly. Determining the appropriate process generally starts by a classification of the "core make" iron. A single part that will be manufactured in large quantities is a candidate for a transfer line. A family of many like parts is a candidate for an FMS. A family of fewer like parts could be processed

by a cell or cellular system. When the appropriate
process is not inherently obvious, all alternatives
should be tested using discounted cash flow analysis
(see Chapter 10).

3. *Address quality plan.* Quality management is integrated into the process in a continuous flow factory,
so the process selected must have capability and
must incorporate a control system such as probing,
gauging, or a coordinate measuring machine.

4. *Develop systems plan.* The systems plan is a joint
effort between internal systems people and vendors.
The plan outlines hardware and software needs, as
well as corporate standards and specifications
which must be met by vendors to assure area computer integration. The systems plan also projects
the cost of hardware, software, the communication
network, and integrating factory systems at the
area and host levels.

5. *Select an integrator.* A key part of the systems plan
is selection of an integrator to link process computers, material handling systems, and sequencers at
the area level. There is a tendency to delay integration because it is a complex job and the machinery
can function without it. However, the full benefits of
modernization cannot be achieved without integration, because it is the area computer that causes
flow. It is therefore best to plan integration early in
the process and fund it as part of the total capital
investment.

D. Rearrangement Plan

In many modernization projects, up to 99% of total factory
floor space must be changed to accommodate the new layout.
The cost and disruption associated with this change can be
minimized with proper planning. Rearrangement can happen
in a step-by-step manner, without interrupting production,
as long as sufficient space has been identified and freed up

during the prelude to modernization. Other keys to efficient rearrangement:

1. *Assure adequate buffer inventory* to make all moves without interrupting assembly supplies.
2. *Use PERT charts* to outline all steps in the process and the sequence of moves.
3. *Make moves during vacations and long weekends.*
4. *Use professional movers* to increase the speed of a large move.
5. *Communicate* details of the move to all people involved (and share a factory model if possible) so everyone understands the task to be accomplished.

E. Concept Reviews

As detail processing gets underway, it is important to conduct concept reviews with the executive architect. (See approval process, Chapter 10). Holding these informal working sessions early in the process helps assure that the project is viable and supports the corporate vision. Concept reviews also help speed the final approval process and implementation of the project.

F. Financial Models

As concept reviews are completed and agreement is reached on the vision, preliminary financial models can be developed for each minifactory. These models project capital needs, start up costs, benefits ("as is versus to be"), impact on cash flow, and the expected return on investment and return on assets. The models can be consolidated to determine the impact on total plant costs.

G. Plant Financial Models

Plant financial models can then be consolidated to assess the impact on the corporation and acquire endorsement by the top executives and directors. A critical decision must be made as to whether the entire manufacturing base will be modernized simultaneously (the "big bang" approach, see Chapter 2) or whether modernization will take place one plant at a time.

H. Minifactory Proposals

As projects are finalized with firm cost and committed benefit numbers, they can be presented for approval. The checklists from Chapter 10 are useful in the approval process. Things to watch for during the review process:

1. *Assembly is included in the proposal.* Some plants leave assembly out of the proposal, believing that because assembly labor cost is small relative to total costs, an investment to improve assembly is not cost effective. Although the numbers may support that conclusion, nearly all the successful plants we visited concentrated on improving assembly first. There are two key reasons why it makes sense to improve assembly. First, consumption of parts in assembly drives production or commands the flow. By improving the flow commander, we provide a good regulator for the entire factory. Second, improving assembly is the fastest way to improve product quality. Often more than 50% of all defects are assembly related. These defects can be prevented and eliminated with flexible assembly techniques.

2. *Major principles are not violated.* The architect of modernization is wise not to impose solutions on the organization. However, he or she must be strongly committed to defending the vision and its principles. When proposals that violate the vision and principles are submitted (for example, using a central receiving area rather than point-of-use delivery or designing an MRP II-driven factory), it is the responsibility of the architect to convince the proposal developers to change. Normally, people violate the principles because they do not understand them. Constant communication and ongoing educa-

tion will help educate all employees about the vision and principles and assure that all proposals are based upon them.

3. *The best suppliers have been chosen.* It is human nature to choose suppliers with whom we have worked in the past. But a major modernization project may require establishing relationships with new suppliers. Early in the process it is a good idea for proposal developers to put together a list of suppliers who will be contacted about the project. The architect may want to suggest other names, based on experience with and visits to other companies. It is best to agree on supplier candidates before a proposal is completed out of respect for the hard work that is required of all parties involved in developing a proposal. During the approval process, the developers should be prepared to justify why a particular supplier was chosen and why others were eliminated. The architect should not impose a choice on the organization, but should assure that all viable candidates have been considered. When planning a complex process like an advanced FMS, it might be necessary to enter into partnerships with suppliers. There are often too many unknowns to establish a firm price. In addition, the product design may need to be amended as the process matures and a dual effort on process and design may be needed for manufacturability. Good partnerships are beneficial to all parties and should be encouraged with competitive vendors who have been tested on catalog machinery and have a good history on pricing.

I. Employee Preparation

This is a crucial step in the conversion process and can be the determining factor in whether the project is a success or a failure. The topic will be covered in more detail later in the chapter, but key steps in this process are:

1. *Communicate as openly and as frequently as possible* about the modernization vision, the principles upon which it is based, and the status of implementation. Face-to-face communication in small groups is normally most effective, but other vehicles such as newsletters, videos, electronic mail, bulletin boards, and all employee meetings should be used extensively. The goal is to answer questions, eliminate fears, and generate support for change. The major concern most employees have is job security. There is no easy way to address this issue because one of the objectives of modernization is, in fact, productivity improvement through employment reduction. Companies that have been through modernization and employment reduction offer these recommendations:
   - Be honest with employees about the future.
   - Develop a plan for reducing the workforce in the most humane way possible (capitalizing on attrition, offering generous severance packages, providing job placement assistance).
   - Remind employees that jobs remaining after a modernization project will be more secure because the company will be more competitive.

2. *Invest in focused, specific training.* An investment in employee training is essential to get maximum benefits from the investment in machinery and systems. Experience indicates that intensive training specific to the new process (cellular system or FMS) is more effective than general training in math, computers, or literacy. General training is necessary, but it can be delayed during modernization. Specific job-related training is most effective when it is timed to coincide as closely as possible with the delivery and installation of the new machines and systems. In some cases, it may be necessary to have a machining center delivered in advance of the FMS

for training on perishable tools and programming. Another option is building a simulator for training on complex gantry machines. Operators and maintenance people can learn by participating in the assembly of the FMS at the supplier's site. Suppliers should also be willing to help with training classes, manuals, and documentation. Extra manuals should be purchased to allow employees to study them at home. It is rewarding to see how hard employees will work to be successful in their new jobs.

3. *Develop a plan to assure job stability.* High turnover is a threat to the effectiveness of a continuous flow factory. To realize the best return on an investment in machines, systems, and training, it is necessary to maintain job stability. A list of critical jobs should be developed, along with a plan for assuring stability on those jobs. Stability can be gained by raising job classifications and pay rates on critical jobs, and by exploring contractual agreements.

4. *Develop a plan for minimizing hardship to surplus employees.* One of the most critical and sensitive aspects of modernization is employment reduction. To be successful, a modernization program must reduce the size of the labor force, but obviously, no company wants to inflict hardship upon employees. It is the responsibility of top management to develop a plan for surplus employees. The plan could include transferring employees to new activities, retirement incentives, separation packages, and other measures that minimize hardship to employees and their families.

J. Install and Ramp Up Equipment

Installation and ramp up are often characterized by disputes between the purchaser and vendor. Following are suggestions to minimize conflict during implementation and shorten the overall process:

1. *Appoint a champion for each large cell or FMS.* This may or may not be the person who planned the project, but it should be someone with knowledge, experience, expertise, and enthusiasm. He or she must be committed to the project and able to resolve ramp up problems one at a time.

2. *Hold monthly meetings with suppliers* to determine what needs to be done and who will pay for it.

3. *Involve purchasing people* in any discussions with suppliers that affect supplier costs.

4. *Budget at least 10% contingency* in the capital investment to fund company-caused mistakes. This 10% can help fund necessary additions to a project that has been budgeted too tightly. If a 10% addition has a significant negative effect on the return on investment, the project is probably not good enough anyway. And if the contingency is not used, the project will have an even better return.

5. *Make sure the supplier budgets some contingency* for its own mistakes.

6. *Avoid legal disputes.* Very often, implementation delays are caused by financial conflicts, not technical problems. When a budget problem arises, it is best to avoid a legal dispute. Lawyers cannot get equipment up and running. Only the company and the supplier—working together—can accomplish that. The most effective way to resolve problems is for senior executives from the company and the vendor to meet, assess the situation, analyze the root cause of the problem, and develop a mutually beneficial solution.

II. KEY STEPS IN THE CONVERSION PROCESS— A "HOW TO" APPROACH

The following section gives additional guidelines for completing some of the most critical steps in the conversion process. This "how to" approach is based on the experiences of planners and implementers from factories around the world.

A. How To Simplify Processes

1. Identify the most labor intensive administrative processes.
2. In each area, target five processes with the highest potential for labor savings.
3. Dedicate a conference room for work simplification sessions.
4. Involve employees in documenting existing processes. Use flow charting or other techniques.
5. Engage in participative work simplification sessions with employees. Solicit proposals for streamlining paper routing, reducing paperwork, simplifying administrative procedures, etc.
6. To encourage participation, reward employees who contribute process improvement suggestions.
7. Evaluate feasibility of mechanizing simplified processes.
8. Finalize new process proposal.
9. Present new process to affected areas. Modify as needed based on employee input.
10. Train employees.
11. Implement new process.
12. Monitor progress.
13. Amend office layout to accommodate new process.
14. Reassign surplus employees.

See Chapter 1 for information on process simplification.

B. How To Design Plant Closing Study

1. Identify excess capacity by product and by facility.
2. List candidates for consolidation based on:
   • Low utilization

• High fixed (period) costs
• Obsolete processes
• Practicality of moving production to another location

3. Use discounted cash flow analysis to assess impact of each alternative on total plant costs. Evaluate:
   • Cost (Net Present Value—NPV) of consolidating Plant A production into Plant B
   • Cost (NPV) of consolidating Plant B production into Plant A
4. Assess other factors including:
   • Ability to consolidate Plant A into Plant B
   • Ability to consolidate Plant B into Plant A
   • Effect on people—mobility
   • Effect on communities
   • Availability of labor in each location
   • Wage and labor relations trends in each location
   • Quality achievements in each location
5. Pursue only high-return (NPV) alternatives, providing for contingencies.

C. How To Design a Good Layout for Flow

1. Identify final location where material will be utilized (assembly store, subassembly, assembly line).
2. Assure that processes move material step-by-step to final location.
3. Make the last step of the process adjacent to the final location where material is needed.
4. Identify locations where rough material is fed to the process. Place all process feeding points on same line.
5. Locate rough store sequencer and supplier delivery points or docks adjacent to the feeding line.
6. Locate supplier delivery docks close to the point of use (rough store above assembly store, etc.).

7. Let suppliers' trucks deliver material to multiple points of use.
8. Manage weight and distance by minimizing travel of heavy material.
9. Design heavy-part processes to exit close to final location.

See Chapter 5 for more discussion of layout principles.

D. How To Schedule a Continuous Flow Factory

1. Control all material movement by computer, regardless of whether the material handling is mechanized or not.
2. If fork trucks are used, monitor the sequence and timing of manual moves by computer.
3. Use intelligent sequencer and algorithm (Figure 4) to initiate demand for material retrieval, movement, and delivery within desirable window of time.
4. Select modular software (Chapter 7) for the factory execution system. Use purchased software when possible.
5. Use material movement software to manage:
   • Movement within FMS or transfer line (controlled by vendor software)
   • Movement inside minifactory (controlled by area computer)
   • Movement between minifactories (controlled by area or dedicated computer)
6. Use sequencer intelligence algorithm (Figure 4) to generate pull trigger signal to internal and external suppliers and ensure just-in-time replenishment.
7. Use a combination of sequencer intelligence and material movement software to perform scheduling.
8. Activate processes to machine part loads in the order in which they arrive (first in, first out processing).

9. Use MRP II to supervise the pull-trigger signal.
    Resolve griefs daily, making corrections to MRP II.

See Chapter 3 for more discussion of scheduling a continuous
flow factory.

E. **How To Apply Quality Principles**

Engineering
1. Assure critical dimensions are identified on design
    and are within realistic tolerance.
2. Do not ask for more precision than needed.
3. Issue company policy for zero defects on critical
    dimensions.
4. Use simultaneous engineering of the design and
    process to assure process capability.

Processing
1. Design a quality plan by process to assure all tools
    are in place and Statistical Process Control (SPC) is
    defined:
    • Gauge list, gauging fixtures, machine probes, air
      gauge, coordinated measuring machine (CMM)
    • Measuring and sampling procedure or on-
      machine inspection procedure
    • Charting of critical dimensions, deviations, use of
      CPK measurements
    • Use of computer aided SPC for efficiency and
      record keeping

Operator
1. Make process control the responsibility of the opera-
    tor.
2. Have quality training and certification of operators.

Assembler
1. Assure all material is available before starting
    assembly.

2. Assure all assembly tools are in place and ready to be used.
3. Develop a process for maintaining tool quality and calibrating air tools and wrenches.
4. Design ergonomic assembly stations with lifting devices or platforms to access product.
5. Conduct assembler training and certify assemblers.
6. Have assembler's name or number stamped on major assemblies to increase accountability for quality.
7. Avoid the paper mill generated by inspectors writing rework tickets.
8. Use informal feedback from testing area to take immediate corrective action at the work station.

See Chapter 3 for more discussion of quality principles.

F. How To Implement a System Plan

General
1. Create a computer integrated manufacturing (CIM) task force by regrouping available talent under the authority of the executive in charge of modernization.
2. Establish realistic integration objectives. For example, integrate engineering data, Manufacturing Resource Planning II (MRP II), and factory floor systems.
3. Establish standard for data base management software (e.g., Oracle™).
4. Build up the operational data base to serve existing systems.
5. Suppress all individual data bases when ready.
6. Establish standards for MRP II, CAD/CAM, engineering work stations, and three-dimensional software.

Hardware

1. Establish standard (IBM or other) for host computer (Levels 3 and 4).
2. Establish standard (DEC or other) for area computer (Level 2).
3. Use vendor hardware for Level 1 computing to avoid changing vendor standards.
4. If vendor hardware is not available, use DEC for Level 1.
5. Establish standard (IBM or other) for personal computers and terminals for the shop floor. All personal computers and terminals will need to be integrated with Level 1 and 2 systems.

Software

1. Get consensus on basic modular software and vendors based on functionality and services.
2. Select off-the-shelf material movement software (e.g., DEC, General Electric, EDS).
3. Select off-the-shelf generic cell controller software (e.g., British Aerospace).
4. Use pull trigger at part level to schedule the factory in harmony with MRP II and the Figure 4 algorithm.
5. Purchase or develop internally an MRP II software (e.g., IBM COPIC).
6. Segregate parts into procurement families (e.g., 40% delivered just-in-time by suppliers, 20% minimum inventory, 30% pull trigger, 10% MRP II).
7. Design corporate specifications for factory integration. Have all machine vendors comply with corporate specifications and the Manufacturing Automation Protocol (MAP). This will define information flow and discipline transaction formats to permit integration.

8. Select an integrator for all manufacturing systems
    at the area level (Level 2). Involve integrator early
    in the process.

See Chapter 7 for more systems information.

III. MOST COMMON MISTAKES

Many lessons have been learned by modernizing 18 factories
at Caterpillar Inc. and consulting to companies undergoing
modernization. This section outlines some of the most common mistakes companies have made during major modernization programs. Avoiding these mistakes helps speed
implementation and improves the overall return on the
investment. Most mistakes are made because people do not
fully understand the principles of continuous flow manufacturing. As a result, their vision of modernization is narrow
and incomplete. Key mistakes include:

A. Flow Is Compromised

The fundamental source of performance and capital savings
in a continuous flow factory is a layout with good flow. If it is
not possible to develop a good layout, modernization will
likely be a waste of capital. A frequent mistake is trying to
reduce the cost and disruption of rearrangement by putting
new machinery wherever there is room in the building. This
may reduce direct labor costs and improve quality, but it will
generate a poor return on investment. To achieve all the
benefits of continuous flow manufacturing, a comprehensive
rearrangement is critical. Major machines and systems must
be located in the right position relative to one another and to
assembly. A block layout with minifactories is fundamental.
Creating smaller, more manageable entities makes it easier
to select, install, ramp up, and integrate the new machines
and systems. Minifactories also permit more accurate monitoring of the investment.

B. The Need for Flexible Equipment

Flexibility is not an option in a continuous flow factory—it is an imperative. To achieve the benefits of continuous flow manufacturing, machines must be capable of processing a variety of parts in any order from start to finish.

The example below shows a traditional factory without flexible equipment. Nine different types of machines (A, B, C, D, E, W, X, Y, Z) are required to process five parts. Some machines (A, D, W) are heavily burdened and produce a bottleneck in the flow. Other machines (C, E, X, Y, Z) are underutilized and stand idle for long periods of time. Parts are processed one operation at a time and must wait between operations. Flow is intermittent and there is much potential for scheduling error.

Traditional Factory

Part 1	A	B		D				Z
Part 2	A			D		W		Y	
Part 3	A			D	E	W		
Part 4	A			D		W		
Part 5	A		C	D	E	W	X	Z

The next chart shows these same five parts being processed in a continuous flow factory with flexible automation. The iron is segregated into three families (small, medium, and large) and processed in one place from start to finish. There is no scheduling and no parking of material between operations.

Flow Factory

Parts 1 and 2	Small part FMS
Parts 3 and 4	Medium part FMS
Part 5	Large part cellular system

Equipment flexibility is essential in a continuous flow factory. It enables the assembly area to draw from the sequencer and the sequencer to replenish itself quickly because the equipment can process all the iron coming to it, in any order, with short lead time.

Some mistakenly believe that flexible equipment is not as productive as one-of-a-kind specialty machines. But the fact is, even though a specialty machine may perform its operation at a high speed, it takes up to 90 days to process a part with stand-alone machines. Flexible machinery can get the job done in 5 days or less.

C. Prelude to Modernization

Failure to complete the prelude to modernization takes its toll on the complete program. Companies that do not rationalize and simplify early in the process cannot reap the full benefits of automation and integration. Some of the most frequent problems with the prelude include the following:

- There is no clear product strategy, so capacity requirements are not known with certainty.
- Surplus equipment is not disposed of.
- Underutilized plants remain open.
- The company elects to continue making parts that could be purchased more economically from low-cost, high-quality suppliers.
- The supplier organization is not reduced to a size that allows implementation of standard packaging, standard bar code identification, and just-in-time delivery to points of use.
- The discipline of simultaneous engineering is not fully utilized.
- Standardization of material codes, sizes, and steel plates does not take place.

D. Internal and External Logistics Plans

Outside of the automotive industry, few companies have developed efficient internal and external logistics plans. By

ignoring logistics or delaying their implementation, a
company sacrifices many benefits. As noted in Chapter 2,
logistics and logistics systems require 20% of the total modernization investment, but cause 40% of the total returns.
Following are some of the major mistakes companies make
regarding logistics:

- Suppliers do not use standard packaging and electronic identification, so it is not practical to deliver to
  the point of use. As a result, a central receiving area
  is still necessary, and there is an accompanying need
  for additional material handling and people.
- Suppliers do not monitor bins and boxes at the
  assembly location and replenish as needed.
- Internal logistics systems (sequencers and computer
  controlled material handling) are delayed. Without
  them, there is no continuous, synchronous flow. The
  cost penalty for this decision is high as more than 80%
  of the investment has been made, but only 60% of the
  benefits can be realized.

E. Production with MRP II or Another Push-Type System

As discussed in Chapters 3, 6, and 7, a push-type scheduling
system based on batches of work orders generates waves of
production in the shop. Problems associated with production
waves include poor machine utilization, interrupted flow,
high in-process inventory costs, excessive material shortages,
and scheduling disruptions. The pull trigger or electronic
Kanban system is superior to MRP II, creating an efficient,
manageable and cost effective flow of materials.

F. Quality

Making quality a separate function, measured by quality
control employees, dilutes accountability for quality. In a
continuous flow factory, quality is the number one priority
and everyone's personal responsibility. Quality improvement

is reflected in every aspect of a continuous flow operation. There are no compromises when it comes to quality. Flow stops for any quality problem. Products are designed for quality manufacture. Processes are capable. Process control is utilized. Quality training and tools are made available to all employees. Everyone is measured and rewarded according to their quality improvement achievements. All employees, suppliers, and minifactories are quality certified.

G. Relationships with Suppliers

In the United States, suppliers of purchased finished goods are often treated as adversaries. Large companies have learned to flex their purchasing muscles by mandating price cuts and dictating rigid supplier practices. In Japan, however, it is more common for manufacturers to perceive suppliers as partners. All parties work together to develop mutually beneficial solutions to their cost and quality challenges. Suppliers in Japan are considered to be extensions of the factory. Company and supplier employees work in teams, focused on cutting costs—not price. It is not unusual for Japanese companies to ask suppliers, "What can be done to make our parts easier to manufacture? What changes do you suggest in our design? How can we improve our packaging? Can our procurement methods be streamlined?" Working together to resolve these issues and lower the supplier's costs ultimately improves profitability for all parties. In the United States, relationships with machine tool vendors can be adversarial. During the concept phase, companies often demand technological sophistication beyond the level a supplier can offer. Suppliers commit to delivering the sophisticated product because they want the business. The result is a prototype machine, with difficult ramp up due to unnecessary gadgets or systems.

It is a mistake to demand too much technology from suppliers. It is also a mistake to impose hardware or software standards on a supplier, except in the case of perishable tooling. Other mistakes can be avoided if, during ramp up, the plant manager communicates regularly with a top executive in the supplier organization, and if all parties agree upfront to pay for the problems for which they are responsible.

H. Automation

Keeping automation and systems simple is a fundamental principle of continuous flow manufacturing. There is a price to pay for complexity—longer ramp up time, process instability, and erosion of return on investment. Although unnecessary complexity should be avoided, there is also danger in being too conservative. If a company limits itself to primitive cells and cellular systems, it may not realize sufficient cost reduction. When competition has a big cost advantage and a quantum improvement is needed, FMS technology will likely be necessary. If high utilization is needed, a group of linked FMSs may be required. The key is to use proven technology, stay within the supplier's capabilities, and avoid both extremes—unnecessary complexity as well as simplistic machinery that will not produce enough benefits.

I. Piloting Continuous Flow Manufacturing Concept

Some companies want to prove the value of continuous flow manufacturing by conducting a small-scale pilot in one area of a plant. Piloting continuous flow concepts is difficult, if not impossible. A flow factory requires a critical mass—an entire plant producing one or more families of product. The entire environment must be changed to incorporate just-in-time logistics, supplier delivery to point of use, packaging standards, electronic identification, sequencing, automated material handling, and all the systems that support these concepts. A flow factory also requires a new organizational design and the elimination of a separate quality function, scheduling, work standards, and others. It is not practical to create such an environment in a single section of a plant.

The only aspect of continuous flow manufacturing that can be piloted is flexible machinery, and as we have discussed the full benefits of flexible automation cannot be realized without the automated material handling and area computer integration. It is a mistake to attempt a pilot on a small-scale basis. A better approach is to visit companies that have modernized, review their principles, and look at their results.

J. Overly Optimistic Numbers

Two areas of a modernization project are often underestimated: ramp up time and start up costs. Both can have a negative effect on ROI. To reduce ramp up time and start up costs:

- A 5 to 10% contingency should be budgeted to cover the cost of unknown or unexpected problems and speed up implementation of solutions.
- Complexity should be minimized.
- Plant management should monitor implementation closely.
- A two-way partnership should be established between the company and the vendor.

K. Basis for Project Approval

Although it is vital to analyze the financial impact of a modernization project, the decision to approve a project should not be based solely on numbers. A good rule of thumb is to approve a proposal if it:

- Is based on the principles of continuous flow manufacturing
- Meets all the criteria outlined in the checklists in Chapter 10
- Delivers a good return on investment, despite a 5 to 10% contingency

Sometimes accountants will want to analyze alternatives to a complete proposal. They will ask to complete a discounted cash flow analysis on the proposal with and without the sequencer, the automated material handling, or the area systems. The problem with this approach is that it is difficult to link the benefits which will be realized to a specific asset that may be eliminated in an incomplete project. A flow factory is a whole—an indivisible concept. Total benefits can only be realized by applying all the principles and assets. Only a comprehensive modernization project will achieve maximum results, so there is no point in analyzing alternatives to a complete minifactory proposal.

L. The Project Approver's Proposals

Developing a complete modernization proposal can be a lengthy and time consuming process. Frequently, projects have to be revised and recycled in order to comply with the principles of continuous flow manufacturing. If a project has been reworked one or more times, some approvers are hesitant to recycle it again. Pressure to move quickly or empathy with all the hard work that has been done may cause the approver to support a project, even though it compromises the principles.

Experience shows that being too flexible during the approval process adds time and cost to the project. Concessions made in the early stages are usually regretted later on. If the principles are firmly adhered to when the proposal is developed, implementation will be faster and easier and the investment will pay a better return. Each principle is important, but the most critical one is layout. Good flow is the key to all gains, so no proposal should be approved unless it includes the best possible layout.

IV. MANAGING CHANGE

Managing change is one of the most challenging aspects of modernization. Even the best managers do not always know how to prepare employees for the changes that will accompany modernization. When people are not well prepared for change, they resist it. Resistance to change can have many negative effects including poor morale, slow implementation, or sabotage. Sabotage is the most difficult effect to detect because it goes underground.

People at different levels of the organization react to change differently. Workers may resist change because they fear the loss of their jobs. However, those who perceive the severity of the economic challenges a company faces may welcome change if it will strengthen their job security by improving competitiveness. Managers at upper levels tend to support change because they understand company strategies, know why change is necessary, and have been part of the decision process. It is middle managers who are often most resistant to change. They not only have less exposure to company strategies than upper managers, but they are also more directly affected by the decision to modernize. They are the doers of the organization; they will be responsible for implementing the change.

The support of middle management, and employees at all other levels, is critical to the success of a modernization project. It is difficult to gain that support when change is forced upon the organization, as shown in Figure 44, where top management hammers a square peg into a round hole. A

Figure 44   Implementing changes I.

Figure 45 Implementing changes II.

more effective approach is shown in Figure 45. Management prepares the organization for change by making the round hole square so the peg can be slipped in easily.

It is difficult to prepare people for change, but it can be accomplished. Following are some suggestions for building support and enthusiasm for modernization throughout an organization.

A. Business Environment

The business environment itself can help people prepare for change. In recent years, people have become increasingly aware of changes in the global marketplace. They have seen their companies attacked by aggressive competitors. They have read about or experienced layoffs and plant closings. They have been challenged to reduce costs and improve quality in their own jobs. They have grown accustomed to changing technology in the factory and office and faster, more frequent new product development cycles.

As the business environment has become more volatile, people have begun to realize that their companies must be more dynamic. They do not always welcome the disruption of change, but common sense tells them that it is inevitable. Fighting change will not be in their own best interests or the best interests of their company.

B. Communication Plan

The importance of communication during modernization cannot be underestimated. It may be necessary to maintain confidentiality during the early stages of strategic planning, but the time will come when the modernization plans will be made public. All too often, companies focus on the legal aspects of communicating to stockholders and the investment community, while minimizing communication to employees. While it is obviously necessary to meet the legal requirements associated with a major investment, it is also critical that employees learn about modernization changes from the company—not from the media.

Communicating openly and frequently to employees is one of the best ways to gain support for modernization. Internal communications can take many forms including:

- Formal presentations or speeches to employees, given by the architect or data gathering team.
- Presentations of approved projects by those who developed them. These sessions should be given to large audiences of the developer's peers and allow time for suggestions and input.
- Presentations by plant or unit managers that communicate specific plans for an individual factory.
- Scale models, plant layouts, artist's renderings, and other displays that show how the plant will look.
- Employee meetings explaining why change is necessary, the expected results, and plans for reducing employment.
- Messages in internal publications, videotapes, electronic mail, and other communication vehicles.

In addition to communicating the modernization vision, other key elements of the communication plan include:

- Information that proves the need for change (customer comments, competitive developments, market position, and current cost levels).
- Evidence that a project of this magnitude can be accomplished (results from other companies that have modernized successfully).
- The importance of teamwork and participation (description of the roles each function plays).
- The advantages of improving competitiveness through a multifunctional approach (modernization), as opposed to a single-functional program (e.g., a zero-defect program sponsored by the quality department, a new research and development initiative supported by engineering, or a new promotion effort launched by marketing).

C. Participation in Proposal Development

Experience indicates that when employees participate in developing a modernization proposal, they are more apt to understand and support it. Involving engineering, manufacturing, marketing, and quality managers early in the process helps prevent functional battles and build a strong multidisciplined base of support for the project.

It is especially important to involve middle managers in proposal development and decision making. The middle management team is large in size and relatively invisible in an organization. If middle managers are negative about a project, they pose a significant threat to its success. Using their talent and experience early in the process helps assure their understanding of and support for the vision. And ultimately, broad participation in proposal development by middle management helps convert a large group of possible spectators or critics into a positive and competent team of implementers. If there is a problem with a particular middle manager, a visit to a site using continuous flow manufacturing may be effective. We have seen a situation in which a department head was opposed to hiring an integrator for a large assembly project. After he and the plant manager visited two factories that had used professional integrators, he changed his mind. The trip was the deciding factor. Talking would not have done it, and forcing the issue would have resulted in alienating a key person.

D. Enlisting Proposal Developers

Proposal developers should be encouraged to present their concepts and plans to one another and to the organization. This process helps educate managers about the basic principles of continuous flow manufacturing. It also provides the management team with a broad perspective of the changes that will take place in the company and most importantly, it serves as a way to build support for the vision. Each developer is, in effect, a sponsor of one aspect of the vision. As more proposals are developed, approved, and communicated, the number of sponsors grows and support for the vision solidifies.

E. Explaining Approved Proposals

As proposals are finalized and approved, they should be simplified and explained to employees. Key areas to communicate include the basic principles of the proposal, the investment required, expected benefits, and the timetable for implementation. Employees should also be told they will have the opportunity to provide input into the design of their work stations and other aspects of the plan. Communicating the key elements of the proposal and soliciting suggestions from employees help strengthen support for and commitment to the plan.

V. SUMMARY

This chapter has outlined a process for converting a traditional factory to a continuous flow factory. It has also discussed "how to" approach critical steps in the process, mistakes to avoid while modernizing, and ways to prepare an organization for change.

The conversion process involves the following major steps:

- Develop and achieve consensus on a vision.
- Begin the prelude to modernization.
- Begin detail processing at the plant level.
- Develop a rearrangement plan.
- Begin concept reviews.
- Develop financial models for each minifactory and consolidate at plant level.
- Consolidate plant financial models at the corporate level.
- Present minifactory proposals to management and board of directors for review and approval.
- Prepare employees for change.
- Install and ramp up equipment.

Critical steps in the process which were explained in more detail include:

- How to simplify processes
- How to design a plant closing study
- How to design a good layout for flow
- How to schedule a continuous flow factory
- How to apply quality principles
- How to implement a system plan

Following are important mistakes to avoid:

- Flow is compromised in the layout.
- The need for flexible equipment is not understood.
- The prelude to modernization has not been seriously implemented.
- Internal and external logistics plans are not developed.
- Production is scheduled with the MRP II or some other push-type system.
- Quality is a separate function.

- Relationships with suppliers are adversarial.
- Automation is either too simple or too complex.
- The continuous flow manufacturing concept is "tested" with a pilot project.
- The numbers are overly optimistic.
- Project approval is based on numbers only.
- The project approver is too flexible and approves proposals that compromise the principles.

To prepare an organization for change:

- Take advantage of a changing business environment.
- Implement a comprehensive communication plan.
- Encourage broad participation in proposal development.
- Enlist the help of proposal developers to build support for the vision.
- Explain approved proposals to employees.

Conclusion

I. CRITICAL SUCCESS FACTORS

During visits to more than 100 factories worldwide and discussions with executives and consultants who have implemented the continuous flow philosophy, five factors were identified as critical to the success of a modernization project.

A. Comprehensive Strategy

To achieve the full benefits of modernization, the strategy cannot be limited to one or more aspects of continuous flow manufacturing, such as automation or just-in-time delivery. It takes a comprehensive strategy to reduce total plant costs by 20% or more and make permanent changes in a company's cost structure. A comprehensive strategy is based on the principles in this book and includes such things as product line rationalization; simultaneous engineering; reduction in number of "core make" parts; changes in supplier practices (standardized packaging, bar coding, point-of-use delivery, third-party logistics); layout for flow; pull trigger scheduling; flexible automation; computerized material handling; internal logistics systems; and area computer integration.

B. Complementary Strategic Initiatives

A modernization program is just one facet of a total corporate strategy. To compound the benefits of modernization, it should be completed in conjunction with other strategic initiatives (off-shore production, rationalization of product line, plant closings, continuous cost reduction, process simplification, reorganization).

C. Multi-Disciplined Approach

Historically, manufacturing companies have been highly functional, with separate departments for research, engineering, purchasing, quality, marketing, manufacturing, finance, information systems, and human resources. Managers in each discipline were searching for the latest developments in their fields. Focused on achieving a high level of sophistication within their respective functions, they began to lose sight of the total enterprise. As a result, many organizations became functionally elegant, but uncoordinated and bureaucratic.

In recent years, companies have begun to break down the barriers between disciplines, changing their organizational structures and pursuing team-based strategies with well defined goals. A multi-functional organization, team-based strategies, and clear goals are vital to the successful implementation of a continuous flow factory. To develop a continuous flow factory and keep it running efficiently, a multi-disciplined effort is required. This effort can be compared to that of a surgical team. During surgery, doctors, nurses, technicians, and other professionals join together, each contributing a unique talent, each following a specific routine, each working with purpose and discipline to perform an important job. The concepts of teamwork, routine, and discipline that are so vital in the operating room are also critical in the factory environment. Without them, the company cannot achieve the total benefits of modernization.

D. A Commitment to Principles

Throughout this book, many references have been made to medium or large corporations. This does not imply that continuous flow manufacturing is only for larger organizations. This vision is as applicable to small companies as it is to large ones. The key to implementing the continuous flow vision in smaller companies is understanding the complete spectrum of principles, then applying those that make sense from a business standpoint.

E. A Social Plan

One of the major dilemmas of creating a continuous flow factory is the employment reduction that results from it. Job security is a serious social issue today, and must be addressed prior to implementing this strategy. It is socially irresponsible to confine our interests to technology without taking a hard look at the impact on people. On the other hand, it is also irresponsible for a company to ignore a competitiveness problem and risk losing an entire business and all the jobs related to it.

This book did not cover the social plan for employees who lose their jobs due to modernization. Such a plan must be developed and should be budgeted as a one-time only cost, included in the project's start up costs. A company that expects to be significantly more competitive after modernization can afford to be more generous with social benefits. This is an argument for proceeding with a full-scope introduction to put the company in the best position to offer assistance to displaced employees.

II. INTO THE FUTURE

Figure 46 shows what the flow factory of the 21st century may look like. The artist's model, courtesy of Fritz Werner, was shown at the Chicago Tool Show in the early 1990s. The model shows an interface between sequencing devices and FMS loading and unloading stations. At the top left, an assembly operation is linked to the assembly sequencer. In this factory, material flows in a continuous and synchronous manner to feed assembly. There is no inventory on the floor. Material movement is monitored by area computers so machining is completed during the appropriate window of time. Material handling is simple, with traditional conveyors. The conveyors are short, due to a good flow layout. Sequencing devices are sized appropriately and located to protect the flow.

This is the way many factories may look and operate in the 21st century. In fact, this particular model is actually in operation today at Caterpillar's Joliet, Illinois, plant. Other similar factories are also operational in the United States, Japan, and Europe. These facilities are unique, yet all have certain things in common:

- The product design is advanced, responsive, and matched with the process.
- Prior to modernization, an aggressive phase of rationalization and simplification took place.
- Automation on the factory floor is flexible, but it is neither too simple nor too complex for the job.
- Area computers integrate processes, material handling, and logistics systems.

As more companies modernize their facilities, they will find that the true competitive edge of continuous flow manufacturing is not the level of automation. It is the rationalization and simplification that takes place prior to modernization

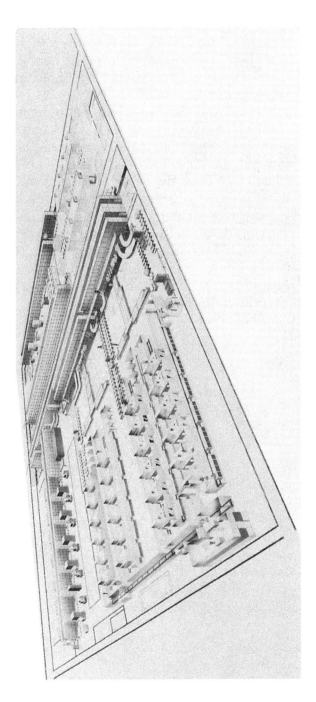

Figure 46   Continuous flow factory of the 21st century.

and the integration that links processes and logistics systems. Many companies are already successfully rationalizing and simplifying their businesses to leverage the gains of modernization. However, only the most advanced are proceeding with integration.

Integration at the area computer level is fundamental to achieving continuous, synchronous flow. It is the only way to assure that processes, sequencing devices, material handling equipment, and pull systems work together in an orderly manner to immediately meet the needs of the assembly line, while minimizing inventory. Integration represents the key to sustained competitive advantage. It is complex and challenging, yet possible today with the area computer at the minifactory level. Companies that seek leadership in the 21st century must be committed to achieving a realistic level of integration with existing technology, while preparing for higher-level integration in the future.

Bibliography

Cost Justification, National Machine Tool Builders Association, 1990. (Based on a study by Ernst & Young.)

Grant, E. L. and R. S. Leavenworth, *Statistical Quality Control*. New York: McGraw-Hill Book Co., 1988.

Investment Justification, Fondulac, WI: Giddings and Lewis, Inc., Nexes™ Automation Division, 1989.

Magaziner, Ira C. and Mark Patinkin, *The Silent War: Business Battles Shaping America's Future*, New York: Vintage Books.

Maschke, Helmut, *Tool Management in a Linked System*, Monchengladbach, Germany: Scharmann GmbH & Co.

259

Glossary

Algorithm. A set of rules for solving a problem in a finite number of steps. An algorithm is typically part of a computer program.

"As is." A set of numbers that portrays a factory situation prior to modernization. "As is" numbers include costs, employment levels, manufacturing space, processing time, inventory levels, and other indicators. "As is" numbers are compared with the "to be" situation to determine the benefits of modernization.

Automated storage and retrieval system (ASRS). A physical structure used for storage in a continuous flow factory. Should be controlled by computer.

Automated storage cell. A group of machines located around an ASRS and used to process family of parts. For each step of the operation, parts are retrieved, processed, and returned to storage.

Capacity. The maximum amount of output a company can produce.

Cellular system. A group of CNC machines, linked with material handling devices and coordinated by computer, that completely processes a family of parts.

Certification. The process of holding operators, minifactories, and suppliers accountable for meeting a set of stringent quality standards.

Champion. An individual with the knowledge, technical skills, enthusiasm, and communications expertise necessary to carry a modernization project from start to finish.

CNC machine. A machine that uses computer programs to control displacement, spindle feed and speed, and tool changes.

Concept review. An opportunity for key people involved in a project to meet and exchange ideas with the modernization architect early in the development process. The objective of such a review is to make sure the project will support the overall vision.

Continuous flow manufacturing. A manufacturing method that permits materials to move rapidly through a plant from the rough-material stage to finished product with little or no interruption. Such a factory is compact, focused, cost effective, and able to adapt immediately to small variations in demand. It is characterized by efficient layout, minimum material handling, minimum inventory, fast response time, and integrated quality management.

Convertible transfer line. A type of transfer line that can be changed to produce another part number. Modification normally allows reuse of most machines and equipment and can be completed within two months.

CPK. A statistical formula that measures quality as a combined function of process capability and process control.

Discounted cash flow (DCF). An accounting method that calculates the current value (or net present value) of benefits that will be accrued in the future.

Equilibrium exchange rate. A currency exchange rate that reflects projected inflation rates in each country and can therefore be used for long-term planning.

First in, first out processing. The ability to process loads of parts in the order of their arrival.

Flexible assembly system. A concept using stationary assembly stalls fed by a kitting system and a computer controlled shuttle to move product.

Flexible carburizer. A heat treat furnace that is capable of managing different carburizing cycles for different loads of parts that have arrived in a random manner. Each load is exposed to different temperature zones and is monitored by a computer that commands furnace rotation and timing of the exposure.

Flexible transfer line. A computer controlled transfer line that can process a family of like parts. Generally made of CNC machines and a few dedicated machines, it has the productivity of a dedicated transfer line, but accepts more variation in part numbers.

Flexible machining system (FMS). An automated, computer-controlled process which is capable of completing a large family of parts from start to finish with no set-up or interruption. Can include tooling migration and quality management and can run unattended.

Flexible welding system. A new factory concept capable of producing a large variety of weldments in a continuous manner. Tacking is performed in stations that are fed by kits of material from the welding sequencer. Welding stations are manual or robotic. Kit and welding assemblies are moved by computer controlled material handling.

Forced flow cell. A Japanese method of processing a family of parts. Similar to an FMS, a forced flow cell is made up of conventional machines set up in a line. Parts are moved manually from one machine to the next, and flow goes in one direction.

Full-scope installation. A FMS which requires all technological advancements including tool delivery, coordinate measuring machine, and computer controlled material handling to be installed in one phase.

Fundamental principles. The primary rules upon which a continuous flow factory is based.

Fuzzy logic. Software logic that can learn from results and apply the knowledge as appropriate.

Gantry robot. A robot that operates on an overhead monorail, moving material or tooling from above a machine.

Griefs. Conflicts between two types of software that monitor and control similar operations. In a continuous flow factory, griefs arise between the execution system (in the area computer) and the MRP II (in the host computer). These conflicts must be reconciled in a timely manner.

Head changer. A device that permits a CNC machine to use highly productive multispindle heads. It also allows the heads to be changed, depending on the configuration of the part in process.

Integration (continuous flow factory). The ability of different information systems to work together, using the same data base, creating a coordinated and synchronized flow of information that drives a coordinated and synchronized flow of material.

Islands of automation. A group of processes or machines within a factory that are linked and controlled by computer and are achieving a synchronous flow of material.

Kanban. A Japanese principle of scheduling production based on replenishing consumed goods. As parts are used, suppliers receive a signal to replenish them.

Kitting. A network of sequencing devices and computer controlled material handling equipment that feeds an assembly or welding station just in time with a kit that contains all the parts necessary to perform the operation.

Layout. A "map" of the factory that allocates space and determines the relative proximity of machines, stores, subassembly areas, assembly areas, material handling devices, offices, aisles, utilities, cafeterias, etc.

Layout block. A layout design based on the minifactory concept in which stores, processes, and machines needed to produce a subassembly are grouped in close proximity.

Make/buy decisions. Decisions that determine whether a company will produce a part internally or buy it from a supplier.

Manufacturability. Property of a part's design that makes it simpler and more cost effective to process without changing its functionality.

Manufacturing Resource Planning II (MRP II). Software widely used by industries to plan material procurement. (II indicates the newest generation of software containing more functionalities like factory scheduling).

Manufacturing velocity. The speed with which a factory processes parts and assembles products.

Merry-go-round FMS. The simplest existing FMS configuration with off-the-shelf machinery and systems. Consists of CNC machines controlled by an FMS computer and laid out on one side of a loop conveyor with loading and unloading stations.

Modular, transportable software. Software with generic capability and functionality to serve a segment of the vision. It can be integrated with other software and can be used by different manufacturing units.

Net capital outlay. A more realistic assessment of a modernization project's capital requirements. Considers the impact of the sale of obsolete equipment and ordinary replacement capital (capital that would be spent even if there were no modernization). The formula is as follows: Net capital outlay = modernization capital − proceeds from sale of obsolete equipment − ordinary replacement capital.

Office cells. A method of streamlining an administrative process by grouping employees from all functions involved in the process in close proximity.

Operating rules. Basic rules that create an environment in which a continuous flow factory will flourish.

Period costs. Costs that remain fixed regardless of volume. Include overhead costs necessary to maintain manufacturing base.

Point-of-use delivery. Logistics concept and layout that require suppliers to transport and deliver materials to places in the factory where they will actually be used.

Prelude to modernization. The first two steps in a modernization plan—rationalize and simplify. Taking these actions compounds the gains realized in the last two steps of modernization—automate and integrate.

Primitive cell. A group of traditional machine tools, activated and coordinated manually, that processes a family of parts.

Process capability. When programmed to produce a given dimension, a machine will respond with minor variations due to its built-in level of precision. Process capability is the statistical variability around the dimension required for a number of parts produced.

Process control. For a machine to have good capability, its level of precision must permit it to meet the dimensional tolerance of a design. If the aim of the capability is right on target, the machine is set appropriately and the process is "in control."

Pull trigger concept. A scheduling methodology which signals internal and external suppliers to replenish parts as they are used in assembly. It is an "electronic Kanban."

Rearrangement. Process of repositioning new and existing equipment to create minifactories that produce unique assemblies. All aspects of a minifactory are located in a manner that facilitates flow and minimizes the use of material handling equipment.

Replacement capital. Capital that must be spent to replace worn-out equipment and keep a factory running with the existing layout, concept, and manufacturing method.

Sequencer. An "intelligent" store which may or may not be mechanized, but is controlled by an area computer. It synchronizes and regulates material flow. It can disburse rough material to processes or finished parts to assembly. It replenishes itself by sending a pull trigger signal to internal and external suppliers for just-in-time delivery of material.

Simultaneous engineering. Discipline that permits the design of a product and process at the same time. Benefits include faster new product development; improved manufacturability; and simpler, lower cost designs.

Statistical Process Control (SPC). Methodology to manage quality by measuring process capability and process control.

Synchronous flow. The ability of materials to arrive at a process or at assembly on time, in the right order, and in the right quantity. Synchronous flow requires computer controlled material handling, sequencers, and the pull trigger concept.

"To be." A set of numbers that portrays a factory situation after modernization. "To be" numbers include costs, employment levels, manufacturing space, processing time, inventory levels, and other indicators. All numbers are compared with the "as is" situation to determine the benefits of modernization.

Tool compensation. The ability of a CNC machine to modify the zero baseline of the program which commands machine displacement. This modification is called tool compensation when it adjusts for tool wear or an error in tool setting.

Tool delivery. A method of purging unnecessary tooling and delivering new tooling to an FMS to permit the processing of a new load of parts. Automated tool migration requires the use of tool buffers, tool racks, and tool loading and unloading robots.

Tool management. The practice of replacing tooling after a predetermined number of cuts to avoid unacceptable wear.

Variable costs. Costs that vary in proportion to production volume, such as direct material, direct labor, etc.

Vision. A blueprint for a comprehensive modernization project that is based on all fundamental principles and operating rules.

War room. Room dedicated to a large modernization project. Can include signs, photographs, charts, layouts, scale models, and other visual aids that communicate the vision and fundamental principles. The room is an important aspect of a communications strategy, as well as a central location for concept reviews, project work, approval sessions, and other modernization meetings.

Index

Acquisition strategy, 7
Administrative processes, simplifying, 10–11
Algorithm, intelligent store, 44–46
Approving modernization investment, 204, 243–244
Area computer
  control of material handling, 41
  factory execution system, 49
  Level 2 computing, 151

[Area computer]
  simplification of integration, 158
  software, 153–154
"As is"/"to be"
  defined, 183
  use in financial justification, 196
Assembly, flexible
  concepts, 94–100
  defined, 66
  layout, 97

Assessing modernization projects, 204–212
Automated guided vehicle (AGV), 67–68, 95
Automated material handling, benefits, 28–29
Automated storage and retrieval system (ASRS), 42
Automated storage cells, 68
Automation
 cellular systems, 76–83
 challenges, 103–126
 characteristics of machinery, 40
 common mistakes, 242
 criticisms, 26–28
 element of modernization vision, 17
 flexible assembly and welding systems, 94–100
 flexible machining system (FMS), 80–94
 gains, 175–180
 manufacturing cells, 66–73
 transfer lines, 73–75
 types, 66

Bar code technology, 56–57
Basic principles of continuous flow manufacturing, 36–41
"Big bang" implementation, 31–32
British Aerospace tool delivery system, 92–93
Brown Bowery Corporation tool delivery system, 92
Buffer, inventory, 39
Business operating rules, 16, 33, 163

Business unit reorganization, 12–14

CAD/CAM
 computing islands, 161
 standardization, 146
Capacity, excess, 21
Capital, 197
Carburizer, flexible, 104–107
Cash flow, 199–202
Caterpillar Inc.
 area computer integration, 151
 automated tool delivery, 88
 flexible assembly system, 97–100
 flexible welding system, 100
 stationary assembly, 94–95
Cell
 checklist, 208
 controller software, 153
 manufacturing, 66–73
 utilization, 71
Cellular systems
 checklist, 208
 defined, 66
 examples, 76–80
 versus FMS, 81–83
Central receiving, disadvantages, 119
Certification, supplier, 56
Challenges
 of automation, 103–126
 of creating continuous flow factory, 217
Change, managing, 244–249
Checklists for assessing proposals, 204–212

Chemical processing, location, 120
Communication, importance, 247–248
Computer aided design/manufacturing (*see* CAD/CAM)
Computer controlled material handling, importance, 41
Computing islands, 161–162
Concept review, 204
Consolidation
   of suppliers, 24
   of traffic, 57
Continuous flow factory
   advantages, 136
   basic principles, 36–41
   conversion process, 217–230
   critical success factors, 253–256
   human resource principles, 58–61
   material handling and scheduling principles, 41–51
   quality principles, 51–56
   supplier principles, 56–58
   versus traditional factory, 127–139
   vision, 35–63
Control point, continuous flow versus traditional factory, 131
Conversion process, 217–230
Convertible transfer line, 74–75
Coordinated measuring machine (CMM), 56, 87, 123–124
Corporate-level financial justification, 199–203

Cost management, continuous flow versus traditional factory, 135
Cost reduction
   continuous cost reduction strategy, 9–10, 15–16
   period costs, 189–191
   potential for reduction through modernization, 14, 191–192
   ultimate cost killer, 16
   variable costs, 187–189
Cost structure, 185–186
CPK, 55
Critical success factors, 253–257

Data base systems, 143–144
Decentralized heat treat, 120–121
Dedicated transfer line, 73
Defensive strategies
   continuous cost reduction, 9–10, 15
   diversification, 11, 15
   emigration, 8, 15
   versus offensive strategies, 6, 15
   process simplification, 10–11, 15
   strategic alliance, 7–8, 15
Delivery to point of use, 57, 119, 135
Digital Equipment Corporation
   DECNET, 161
   FMS computer, 150
Direct labor costs, 178
Discounted cash flow, 200–202
Diversification, 11–18, 15

Ease of management, 136, 180
Emigration, 8, 15
Employees
  human resource principles,
    58–61, 172
  involvement in process
    simplification, 11
  managing change, 244–249
  preparing for continuous
    flow, 227–229
  pride, 58–60, 137
Employment levels, continuous
    flow versus traditional
    factory, 137
Engineering
  simultaneous, 25–26
  work stations, 147
Equilibrium exchange rate, 24
Equipment
  checklists, 210–211
  need for flexibility, 238
  selection, 168–169
Ernst & Young discounted cash
    flow, 202
Excess capacity, 27

Financial justification
  common mistakes, 243–244
  components of process, 195–
    199
  corporate-level, 199–203
  plant-level, 203
First in, first out processing,
    43–44
Flexible assembly
  concepts, 94–100
  defined, 66
  key rules and principles, 169

[Flexible assembly]
  layout, 97–100
Flexible automation, 40, 167–
    168
Flexible carburizer, 104–107
Flexible machining system
    (FMS)
  automated parts logistics, 85
  checklist, 209
  full-scale, 80
  full-scope introduction, 93–94
  higher technology, 84–93
  integrated washing and mea-
    surement, 86
  internal scheduling, 86
  managing complexity, 83
  merry-go-round, 83
  selection criteria, 81–83
  software, 152
  tooling issues, 87–93
  versus cellular systems, 81–
    83
Flexible transfer line, 74–75
Flexible welding, 66, 100, 169
Flow
  common mistakes, 237
  maintaining, 39
  obstacles to, 116
  protecting, 119
  pulling, 37–39
  stopping for quality prob-
    lems, 53
  synchronizing, 40
Forced flow cells, 69
Fritz Werner tool delivery sys-
    tem, 88–89
Fundamental principles, 33–34,
    61–63, 101–102, 126, 163–
    164, 165–172

Gains
consolidation, 183–194
direct labor costs, 178
ease of management, 180
human resources, 180
indirect labor costs, 178
indirect material and
expenses, 178–179
inventory, 179
leveraging the gains, 192
manufacturability, 175
production materials, 179
purchasing and external
logistics, 174
rationalization, 173
responsiveness, 179
strategy, 173
Gantry FMS, 113–115
Gantry machines, 112
Gantry robots, 91
General Electric
stationary assembly, 94
war room, 221
General Motors
MAP, 160
third-party logistics, 58
Giddings & Lewis
cellular system, clutch brake
components, 78–79
cellular system, fluid pump
casing, 76–77
investment justification, 202
utilization study, 71
Griefs, resolution, 48–49

Hardware, Level 1–3, 148–152
Head changer
Huller Hille GmbH, 108, 111

[Head changer]
Mandelli Industriale Spa,
108, 113
Heat treat
decentralized, 120
flexible carburizer, 104–107
Holcroft Rotocarb, 106
sequencer, 122
Higher technology FMS, 84–94
"How to" lists, 230–237
Human resources
gains, 180
managing change, 244–249
preparing for continuous
flow, 227–229
principles, 58–61
Hundred-factory tour, 35, 253
Hyperchannel, 161

IBM
COPIC, 148, 236
DB2, 143
profit center reorganization,
12
Implementation, "big bang" versus incremental, 31–32
Indirect labor costs, 178
Indirect material and expenses,
178
Information systems
checklists, 211–212
computing islands, 161–162
dimensions of, 141
hardware, 148–152
integration, 142–148, 224
network, 160–161
plan, 224, 235
software, 152–160

Ingersoll gantry machine, 113–
114
Integration
building operation data base,
144–146
defined, 142
delaying implementation,
159
element of modernization
vision, 17
full, 142–143
gains, 175–180
impact on cost reduction, 28–
29
key rules and principles,
171–172
preparing for future, 146
realistic, 144
role in continuous flow
factory, 142
Intelligence
algorithm, 44–47
sequencer, 42
Inventory
buffer, 39
costs, 179
as element of financial justification, 198
in process, 135, 136
reduction, 179–180
Investment monitoring, 212–
214

Kanban
electronic, 45–46
Kayaba, forced flow cell, 69–70
Kitting, 96–97
Komatsu, flexible welding, 100
Kubota, stationary assembly, 95

Layout
checklist, 206–207
continuous flow versus traditional factory, 38, 128–129
gains, 175–180
"how to" list, 232
principles, 36–39, 116–124,
167
rearrangement, 36
Lead time, 45, 134
Level 1–3 hardware, 148–152
Leveraging the gains, 192
Load, 45
Local area network (LAN), 150,
160–161
Logic, sequencer, 44–47
Logistics
checklist, 206–207
concepts, 57–58
gains, 174
most common mistakes, 239–
240
plan, 223–224

Make/buy decisions, 22–24
Mandelli Industriale Spa
head changer technology,
108–112
"over" concept, 112–113
Manufacturability, 167, 175
Manufacturing Automation Protocol (MAP), 160
Manufacturing cell, 66–73
Manufacturing Resource Planning II (MRP II)
common mistakes, 240
limitations for scheduling, 47
with pull-trigger execution
system, 48–51

[Manufacturing Resource Planning II (MRP II)]
  standardization, 147
  as planning tool, 48
Marketing strategy, prelude to modernization, 20
Material handling
  principles, 41–42
  software, 152–154
Merger, 7, 15
Merry-go-round FMS, 83
Mezzanine, 121–123
Minifactory
  checklist, 207–211
  concept, 37–38, 128, 196
Mistakes, most common, 237–244
Mitsubishi, flexible welding, 100
Modernization
  defined, 14–15
  elements of, 17–19
  multiplier effect, 30
  potential cost reduction, 14
  prelude to, 19–26, 33
  scope, sequence, and timing, 30–31
  vision, 17–34
Modular software, 154–155
Monitoring investment, 212–216

Net capital outlay, 197
Network, 160–161
Nippon Denso
  forced flow cell, 69–70

Offensive strategies
  versus defensive strategies, 6–7, 16

[Offensive strategies]
  modernization, 14, 16
  reorganization, 12–14, 16
Office cell, 10–11
Operating rules, 16, 33, 163
Operation data base, 144–146
ORACLE, 143
"Over" concept
  defined, 112–113
  Mandelli Industriale Spa, 110–113
  Scharmann, 113

Packaging, 56
Performance indicators, 202–203
Period costs
  defined, 184–185
  depreciation, 191
  general maintenance, 190
  labor, 189
  as percent of cost structure, 186
  repair maintenance, 190
  utilities, 190
Pilot project, difficulties, 242–243
Plant closing, 20–22, 231, 255
Plant host computer, 152
Plant-level financial justification, 203
Point-of-use delivery, 57, 119, 135
"Poka Yoke," 53–54
Prelude to modernization
  capacity planning, 20–22, 221
  checklist, 205

[Prelude to modernization]
  common mistakes, 239
  make/buy decisions, 22–24,
    220
  marketing strategy, 20
  product line rationalization,
    20
  simultaneous engineering,
    25–26, 220
  supplier consolidation, 24–
    25, 220
Process
  ability to carry iron, 118–119
  capability and control, 54–55
  simplification, 10, 15, 231
  start-to-finish, 39
Product design, 20
Production materials costs, 179
Production planning, 20
Profit center reorganization,
  12–14
Pull concept, 37–38
Pull trigger
  logic, 44–47
  use with MRP II, 48–51

Quality
  bureaucracy, 51
  certification, 56
  common mistakes, 240
  CPK, 55
  employee accountability, 52
  "how to" list, 234
  management, 134, 169
  plan, 224
  "Poka Yoke", 53
  principles, 51–56
  process capability and con-
    trol, 54–56

[Quality]
  statistical process control
    (SPC), 56
  stopping the flow, 53, 134
  using coordinate measure-
    ment machine, 56

Random access gantry FMS,
  113–115
Rationalization
  defined, 19–22
  element of modernization
    vision, 17
  gains, 174
  key rules and principles, 166
  marketing, 20
  product design, 20
  production planning, 20–22
Rearrangement, 36, 224–225
Recap of rules and fundamental
    principles, 166–172
Reorganization, 12–14
Replacement capital, 197
Responsiveness of continuous
    flow factory, 179
Return on assets (ROA), 202–
    203
Return on investment (ROI),
    202–203

Safety stock, 45
Scharmann
  FMS network, 150
  "over" concept, 112
  tool management, 91–93
Scheduling
  by area computer, 47, 50
  "how to" list, 233

[Scheduling]
limitations of MRP II, 47
at part number level, 131,
156
principles, 40–47
segregating iron, 146, 160,
208
Sequencer
algorithm, 44–47
defined, 42–43
intelligence, 42
logic, 43
Shot blasting, disadvantages,
119
Sikorsky Helicopter automated
storage cell, 68–69
Simple cell, 66–67
Simplification
administrative processes, 10
element of modernization
vision, 17
"how to" list, 230–237
Simultaneous engineering, 25–
26
SNK of Japan, gantry machine,
113
Software
principles, 154–160
role of, 156
types, 152–155
Split assembly line, 96, 98
Standardization
for integration, 146–148
software, 155–156
Start up costs, 198
Stationary assembly, 94
Statistical process control
(SPC), 56
Strategic alliance, 7, 15
Strategic gains, 173

Strategic options and planning,
6, 15
Supervisor
conflicting priorities in traditional factory, 132–134
role in continuous flow factory, 39, 134
Suppliers
bar coding, 56
certification, 56
common mistakes, 241–242
consolidation, 24
logistics concepts, 57–58
logistics gains, 174
minimizing conflict with, 230
packaging, 56
point-of-use delivery, 57, 119,
135
principles, 56–58
selection, 227
traffic, 57
Synchronous flow, 40

Third-party logistics, 57–58
Three-dimensional design, 147
Timing of modernization
implementation, 30–32,
198–199
Tooling
automated logistics network,
87–88
compensation, 90
cost, 91
delivery, 88–89
management, 87
scheduling, 90
setting, 90
technology, 91

Traditional factory versus con-
tinuous flow, 127–139
Traffic consolidation, 57
Training, 61, 228
Transfer lines
checklist, 209–210
comparison of types, 74–75
Trigger, 43–44, 48
Turnover, 60, 229

Unit of production, continuous
flow versus traditional factory, 131
Utilization, cell versus FMS,
71–73

Variable costs
assembly, 188
burden, 189
defined, 184
labor, 187
logistics, 188
as percent of cost structure,
186

[Variable costs]
production materials, 187
Vendor checklist, 210
Vision
of continuous flow manufacturing, 35–63
developing and achieving
consensus, 218–219
of modernization, 17–34
Vital few, 18, 23
Voest Alpine integration, 151
Volvo stationary assembly, 95

War room, 221
Welding, flexible systems, 66,
100, 169
Work orders, 130, 132

Xerox Ethernet, 150

Yard storage, disadvantages,
120